CALGARY PUBLIC LIBRARY

FEB - - 2007

Wild Orchids of the Prairies and Great Plains Region of North America

University Press of Florida

Florida A&M University, Tallahassee
Florida Atlantic University, Boca Raton
Florida Gulf Coast University, Ft. Myers
Florida International University, Miami
Florida State University, Tallahassee
University of Central Florida, Orlando
University of Florida, Gainesville
University of North Florida, Jacksonville
University of South Florida, Tampa
University of West Florida, Pensacola

Wild Orchids
of the Prairies and Great Plains Region of North America

Paul Martin Brown

with original artwork by Stan Folsom

University Press of Florida
Gainesville
Tallahassee
Tampa
Boca Raton
Pensacola
Orlando
Miami
Jacksonville
Ft. Myers

Copyright 2006 by Paul Martin Brown
Drawings copyright 2006 by Stan Folsom
Printed in China on acid-free paper
All rights reserved

11 10 09 08 07 06   6 5 4 3 2 1

A record of cataloging-in-publication data is available from the Library of Congress.
ISBN 0-8130-2975-9

The University Press of Florida is the scholarly publishing agency for the State University System of Florida, comprising Florida A&M University, Florida Atlantic University, Florida Gulf Coast University, Florida International University, Florida State University, University of Central Florida, University of Florida, University of North Florida, University of South Florida, and University of West Florida.

University Press of Florida
15 Northwest 15th Street
Gainesville, FL 32611-2079
http://www.upf.com

To Ann Malmquist, friend and companion orchidophile

A Reminder

Our wild orchids are a precious resource. For that reason they should never be collected from their native habitats, either for ornament or for home gardens. All orchids grow in association with specific fungi and these fungi are rarely present out of the orchids' original home. Searching for and finding many of these choice botanical treasures is one of the greatest pleasures for both the professional and amateur botanist. Please leave them for others to enjoy as well.

Contents

Foreword ix
Preface xi
List of Abbreviations and Symbols xii

Part 1. Orchids and the Prairies and Great Plains Region of North America

The Prairies and Great Plains Region of North America 3
An Introduction to Orchids 6
Which Orchid Is It? 8
   Using the Key 8
   Key to the Genera 9
   Some Important Notes About . . . 12

Part 2. Wild Orchids of the Prairies and Great Plains Region of North America

   Aplectrum 17
   Calopogon 21
   Calypso 27
   Coeloglossum 31
   Corallorhiza 35
   Cypripedium 53
   Epipactis 71
   Galearis 77
   Goodyera 81
   Gymnadeniopsis 89
   Habenaria 95
   Hexalectris 99
   Isotria 103
   Liparis 109
   Listera 115
   Malaxis 120
   Piperia 125
   Platanthera 128
   Pogonia 171
   Spiranthes 174
   Tipularia 207
   Triphora 211
Bordering Species 214

Part 3. References and Resources

Checklist of the Wild Orchids of the North American Prairies
and Great Plains Region   219
Provincial and State Checklists   227
Some Regional Orchid Statistics   241
Rare, Threatened, and Endangered Species   245
Recent Literature References for New Taxa, Combinations, and Additions   255
Synonyms and Misapplied Names   259
Using Luer   273
Cryptic Species, Species Pairs, and Varietal Pairs   277

Part 4. Orchid Hunting

   1. The Northern Prairie Region   283
   2. Central Mississippi River Valley Region   286
   3. Islands in the Plains   289
   4. The Black Hills   292
   5. Foothills of the Rockies   295
   6. On the Edge of the Ozarks   297
   7. The Eastern Prairie Element   300
   At the Limit   303
   Tips and Trips   306
   What Next?   307

Appendix 1. Distribution Chart   309
Appendix 2. Flowering Time Chart   312
Glossary   317
Bibliography   321
Photo Credits   327
Index   329

Foreword

The orchid family, Orchidaceae, is the largest family of flowering plants, with at least 30,000–35,000 species. While most are found in tropical or subtropical habitats and are for the most part epiphytic, there are many that occur in the temperate zone. Almost all of these are terrestrial. In the Northern Hemisphere, amazingly, even a few grow as far north as the Arctic Circle.

There have been myriad uses for orchids besides their cultivation for personal enjoyment. The use of orchids in India as medicines (ayurvedic medicine) can be traced back to the *Sushrutra Samhita*, which has been dated to about 600 BCE, and may actually go back to the use of medicinal plants mentioned in the *Rig-Veda* (ca. 1500 BCE). In China, orchids have been grown since before the time of Confucius (551–479 BCE), from whom we have the comment "acquaintance with good men was like entering a room full of *Ian*" (or fragrant orchids). By the fifth century BCE the Greeks were using the twinned tubers of *Orchis*, which resemble testicles, to cure testicular problems (an early version of the *Doctrine of Signatures*). In the region that is now Greece and modern Turkey, south to Israel and Egypt, there is a delicacy called *salep* (salep ice cream) that is made from the dried roots of several genera of orchids. In Meso-America both the Maya and the Aztec cultivated vanilla *tlilxochitl* (black flower/black pod) in Aztec (*sisbic* in Mayan), for use as a perfume and as a seasoning in *chocolatl* (chocolate). The Aztecs also used *tzaconhxchitl* (glue flower), which is an excellent glue for wood.

No matter how much we humans like and appreciate orchids for their beautiful and intricate flowers, or for more practical reasons—their use as seasoning (vanilla), food (salep), fiber for weaving, or an ingredient in glue—we humans are also their downfall because of our destruction of their support systems and habitats.

As the global human population increases we put more and more pressure on the land to produce food, paper pulp, lumber, and other needed materials. The resultant changes in land use in the temperate zone and the clear-cutting of tropical rainforests results in many orchids being lost before they have been collected and described. Our present understanding of the needs of orchids has pointed out that the destruction of pollinators (mainly insects) and fungi needed for seed germination and mycorrhizal associations with the roots/rhizomes/tuberoids of adult plants can eradicate orchids just as surely as pulling them up.

There are three comprehensive treatments of North American orchids: *Native Orchids of North America* by Donovan Correll (1950), *The Native Orchids of the United States and Canada excluding Florida* by Carlyle Luer (1975), and the treatment of the family Orchidaceae in *Flora of North America North of Mexico*, volume

26 (2002). In the years between Luer and FNA volume 26, there have been several new species described and numerous nomenclatural changes made as a result of new information and, in some cases, reinterpretation of older data. Also, new information on the biochemical relationships of various plants has resulted in changes. In *Wild Orchids of the Prairies and Great Plains Region*, Paul Martin Brown has used, with few exceptions, the current FNA nomenclature (names) and also has included information on the presently known hybrids, all of which makes this a very useful book to orchidologists and amateurs alike.

Paul Martin Brown has a Master's degree from the University of Massachusetts. In 1995 he founded the North American Native Orchid Alliance and the *North American Native Orchid Journal* and is presently the editor of this journal. Paul and his partner, Stan Folsom, have several recent publications, including *Wild Orchids of the Northeastern United States* (1997), *Wild Orchids of Florida* (2002), *The Wild Orchids of North America, North of Mexico* (2003), *Wild Orchids of the Southeastern United States, North of Peninsular Florida* (2004), and most recently *Wild Orchids of the Canadian Maritimes and Northern Great Lakes Region* and *Wild Orchids of the Pacific Northwest and Canadian Rockies* (2006).

In the *Wild Orchids of the Prairies and Great Plains Region* Paul Martin Brown treats 21 genera, 64 species, 9 varieties, and 8 hybrids. This comprehensive field guide is divided into four sections. The first section includes an introduction to orchids, how they are different from other flowers, and how to use a key to identify them. The second section consists of keys to the species and varieties of the genera, and descriptions of the genera and species. Color and growth forms that have been described in the literature are listed immediately after the geographic range of the species. He also includes complete descriptions of the genus and species, color illustrations, highly useful line drawings by Stan Folsom, and distribution maps in color. The third section includes checklists for the entire region and for the various states and provinces. There is also a lucid discussion of the rare, threatened, and endangered species in the region and current literature references for new taxa, combinations, and additions, as well as synonyms and misapplied names. Information on the use of the Luer books in conjunction with this book is included. The fourth section includes discussions about orchid hunting and the various geographic localities in this region. There are also appendixes dealing with the distribution of orchids by state and with flowering times, which should be of great value for anyone trying to find wild orchids. Additionally there is an excellent glossary, and an index of both common and botanical names. Paul Martin Brown has done an excellent job of taking a huge amount of data and reworking it into a format that is easy to access and use. Finally his commitment to the conservation of valuable and fragile orchid plants echoes throughout the book. With regard to North American native orchids he is becoming the Carlyle Luer of the early twenty-first century.

Lawrence "Larry" K. Magrath

Preface

The prairies and Great Plains region of North America presents boundless opportunities for the native orchid enthusiast to both search for and discover more than seventy species and varieties of orchids growing in the wild. The primary goal of this work is not only to assemble all of the known information on the orchids in this region but also to pique the interest and curiosity of the user to go to just one more roadside or remnant prairie to find even more wild orchids.

A Note about Field Guides

The primary purpose of a field guide is to assist the user in identifying, in this case, the wild orchids of the prairies and Great Plains region of North America. It is not intended to be either an exhaustive treatise on these species, nor the ultimate reference for that area. It is intended to be used primarily in the field and designed for locating information easily while one foot is in the proverbial bog. The photographs have been taken in the field and are intended to illustrate the species, as the user will see them. They are neither studio shots nor great works of art—just good, diagnostic photos that portray the plants in their habitats. The line drawings also have the same goal, of assisting the user in identifying the orchids. Many times line drawings can communicate different aspects of the plants that photographs cannot. The user is always encouraged to consult additional references for more detailed descriptions and regional accounts for individual species. It is always possible to write more on any subject and more certainly could be said about each species, but first and foremost, this is a field guide. By combining the resources of keys, descriptions, photographs, and line drawings the user should be able, with relative simplicity, to identify all the orchids within the range of this book.

This work, with such a large geographic scope, has involved the advice and counsel of many people. Those who facilitated the research and made major contributions to the information gathered within this book include Norris Williams, keeper, and Kent Perkins, manager, of the University of Florida Herbarium, Florida Museum of Natural History, Gainesville; Gustavo Romero, Orchid Herbarium of Oakes Ames, and Judith Warnement, Botany Libraries, Harvard University Herbaria; Holly Carver, University of Iowa Press; Chuck Sheviak, Paul Catling, Scott Stewart, Ann Malmquist, Lorne Heshka, Larry Zettler, Marlin Bowles, Margaret From, Bill Summers, Carl Slaughter, George Johnson, Tom Sampliner, Doug Martin, Caleb Morse, David Ode, Betty Falxa, and Jim Thorson.

The staff at University Press of Florida have continued their usual support, including John Byram and Gillian Hillis, who contributed much-appreciated editorial assistance, without which this book would never have been completed. And, again, my partner Stan Folsom, whose paintings and drawings grace the pages of this book, has been both a source of support and inspiration throughout the entire project. For that I am most grateful.

Abbreviations and Symbols

AOS = American Orchid Society
ca. = about or approximate, in reference to measurements
cm = centimeter
f. = *filius*; son of, or the younger
m = meter
mm = millimeter
nm = nothomorph or nothovariety, indicating an intraspecific hybrid
subsp. = subspecies
var. = variety
× between or preceding a name denotes a hybrid or hybrid combination
* = naturalized
≠ = misapplied name

Publications:

FNA = *Flora of North America* (volume 26, including the *Orchidaceae*)
INPI = *International Plant Names Index*
NANOJ = *North American Native Orchid Journal*

Part 1

Orchids and the Prairies and Great Plains Region of North America

The Prairies and Great Plains Region of North America

For the purpose of this field guide the prairies and Great Plains region of North America is defined in the map below. The contiguous areas comprising this region stretch eastward from the foothills of the Rocky Mountains in Montana, Wyoming, Colorado, and New Mexico to the Mississippi River Valley and range from southern Manitoba and southern Saskatchewan in the north to north central Texas and much of western Louisiana in the south. The treatment of orchid species in the Deep South, albeit very much within the Mississippi River Valley, has been restricted to those species occurring within prairies or prairie-like habitats. For more details, especially on the Louisiana species, the reader is referred to *Wild Orchids of the Southeastern United States* (Brown and Folsom, 2004).

Orchids and the Prairies and Great Plains Region

The broadly defined region shown on the map includes much more than just prairies and plains; within these vast, open lands occur the major river valleys of the

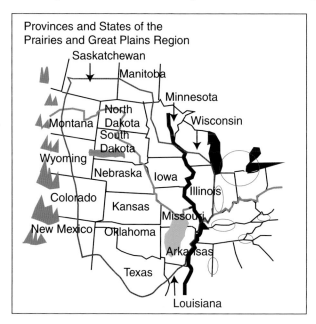

Midwest and isolated, but well-defined, mountain and hill ranges. East of the Mississippi River Valley are numerous prairie islands, most notably in Indiana, Ohio, southern Michigan, and, to a lesser extent, Mississippi and Alabama.

Although no current individual work treats the orchids of the prairies and Great Plains region of North America, portions have been treated, in various ways, in a variety of publications. Larry Magrath's several writings include his dissertation in 1973 and journal articles on the native orchids of Kansas (1972) and Oklahoma (2001). Publication of the Orchidaceae in *Flora of the Prairies and Plains* (Kaul, 1986) and orchid floras of Missouri (Summers, 1987), Arkansas (Slaughter, 1993; Johnson, 2004; Johnson and Slaughter, 2006), Wisconsin (Hapeman, 1996), Indiana (Homoya, 1993), Manitoba (Ames, et al., 2005), and Black Hills (Kravig, 1969) have all been valuable contributions to information on the orchids of this vast region.

Because the prairies and Great Plains extend for more than 1,500 miles north to south, several species are restricted to either geographic extreme or to isolated areas such as the Black Hills of South Dakota and eastern Wyoming. In fact, the Black Hills account for several species records such as the eastern fairy slipper, *Calypso bulbosa* var. *americana*, and slender bog orchis, *Platanthera stricta*, that would not normally be included within a work on the prairies and Great Plains.

Defining both prairie and plain is no simple task. Prairies have so many subspecific categories that calling them simply open grasslands lacking significant woody vegetation is perhaps the most general but simplistic of all descriptions. Four qualities of prairie, whether eastern tallgrass or more western shortgrass, are generally referred to: virgin, reclaimed, degraded, and restored. Several types of prairies may also occur: hill, riparian, savanna, rolling, flatland—just about as many terms as you can conjure up. But the thread that holds them all together is rich, black soil with extensive grasses. True virgin prairie is very hard to come by, but fortunately much that remains is under preservation by a variety of agencies. Some of this land may be as restricted as a narrow strip along a railroad or an old cemetery. Hill and fen prairies are scattered about, especially in Illinois and Wisconsin, and higher floodplains along the rivers are often called riparian prairies. In many of the central prairie states calcareous bluffs occur that also support several prairie species. Oak savanna prairies are a unique habitat found primarily in southern Wisconsin and northern Illinois. They are large expanses of typical prairie grasslands dotted with individual mosscup oaks, *Quercus macrocarpa*. The shade of these oaks offers a somewhat sheltered habitat for several species of orchids. Prairie restoration is widespread and those efforts are slowly reclaiming many areas.

The vast open grasslands further west and south are more correctly addressed as the Great Plains and are usually higher in elevation and drier. Whereas much land in the prairies was plowed for agriculture and development, endless expanses of the plains were used for extensive grazing and, in the north, for agriculture, especially wheat fields. All of these uses have greatly reduced the occurrence of native orchids in these regions.

It could be readily argued that there are only a handful of native orchids found within prairie and plains habitats. Of this dozen or so species four are considered prairie indicators: small white lady's-slipper, *Cypripedium candidum*, eastern prairie fringed orchis, *Platanthera leucophaea*, western prairie fringed orchis, *P. praeclara*, and Great Plains ladies'-tresses, *Spiranthes magnicamporum*. The former two species are also found in fens. Although several other species occur in the prairie/plains habitat they are often more abundant in other woodland or wetland habits. Rather than pick and choose for just species of the open grasslands, all of the orchids occurring from the foothills of the Rockies to the Mississippi River Valley are treated, with more general references to those occurring within the mountains and hills of Missouri and Arkansas.

Regarding all genera found in the prairies and Great Plains region of North America, the recent work in volume 26 of *Flora of North America* (2002) has greatly fine-tuned the identification and distribution of these orchids. The generic treatments in the *Flora* were authored by individuals and teams that were considered authorities on these genera. Although there were some variations in the treatments, specific guidelines were followed by the authors. In respect to distribution the authors verified each state with an herbarium specimen or substantiated record. The orchids were treated at the species, subspecies, and varietal levels. Forma, in general, were not discussed. Some authors also treated hybrids. This compendium of information brings together information that had been fragmented, at best, for many years and also necessitated the completion of many species treatments that had been in progress. Each author or team of authors set their own parameters for delimitation of species, based on their research. As a result, several new species were described and new, or possibly older, names were then used in the *Flora*.

An Introduction to Orchids

Orchids hold a special fascination for many people, perhaps because of their perceived extreme beauty, rarity, and mystery. In actuality, orchids are the largest family of flowering plants on earth with nearly 30,000 species. Although many are exceedingly beautiful, many more are small and dull colored, barely 2 or 3 mm across! There is hardly any place on this planet, other than the Antarctic, that does not have some species of orchid growing within its native flora. Even the oases within the great deserts of the world harbor a few species. In the more northerly climes orchids can be found well above the Arctic Circle. Orchids grow at elevations greater than 15,000 feet as well as within highly developed urban areas of the globe.

In this region, the central part of North America, orchids may not appear to abound throughout, but there are many excellent orchid islands to be found. Some of the rarest species globally are found here. The vast north-south distance accommodates species of both colder northern and warmer Gulf habitats.

Because of the sheer number of orchids in the world there is a considerable degree of morphological diversity within the family. Although all orchids possess certain qualifying characters, their general morphology can be as variable as the imagination. But, viewed closely under a lens, even the tiniest of orchids has the distinctive characteristics that make it an orchid!

Characteristics of the orchid family, the Orchidaceae, are quite simple, despite their diversity. First, they are monocotyledons, or monocots—a major class of the plant kingdom that has a single emerging leaf-like structure when the seed germinates (as opposed to dicotyledons, or dicots, which have two leaf-like structures). Grasses, lilies, and palms are also monocots. Second, orchids have three *sepals*, two *petals*, and a third petal that is modified into a *lip*. This prominent structure is actually a guide for the pollinator—sort of a landing platform that directs the agent of pollination toward the nectary. Third, the stamens and pistil are united into a *column*, a structure unique to the orchids that effects fertilization when the agent of pollination passes by on its way to the nectary. Again, because of the size of the family, there are many pollinators, not just insects. Orchids have been documented to be pollinated by the usual bees, butterflies, wasps, and flies, as well as by hummingbirds, and a few are even rain-assisted in their pollination. Many orchids emit a fragrance at a certain time of day or night to attract

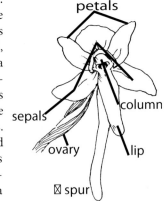

their specific pollinator. Beyond that there are those species that by various means are self-pollinating.

Because of the great number of genera and species, occasionally there are what appear to be exceptions. A few genera have monoecious or dioecious flowers—male and female flowers on separate inflorescences or separate plants. In others, the petals and/or lip may be so modified that they are barely recognizable; in yet others, the flowers never open: they are cleistogamous and self-fertilized in the bud. But all of these exceptions aside, the vast majority of orchid flowers look like orchids!

Within the prairies and Great Plains region of North America, as defined in this field guide, all of the orchids are terrestrial, growing with their roots in the ground and taking in water and nutrients through root hairs and/or swollen stems that are tuber-, bulb-, or corm-like.

In nature, all orchids consume fungi as a food (carbon) source. Each wild orchid has this unique relationship with naturally occurring fungi. While nearly impossible to see with the naked eye, these fungi are essential to the growth and development of any wild orchid. In this relationship, a fungus infects the orchid roots and the fungus is then consumed as a food source, initially prompting seed germination and then sustaining seedling development. As a mature plant, the orchid is capable of producing food via both photosynthesis and the consumption of its root fungus. A number of different fungi may contribute nutrients, growth regulators, vitamins, and moisture at various levels and at different times throughout the orchid's life cycle. It is this special relationship between orchid and fungus that makes native orchids extremely difficult to transplant. In some cases, native orchids that have been moved from their natural location may die within a few years as a result of this disruption of the orchid-fungal association, while in the genus *Cypripedium* damage to the roots often causes death.

The leaves of orchids are nearly as diverse as the flowers. As is characteristic of all monocots, they have parallel primary veins. They may be long and slender, or grass-like, or round and fat, or hard and leathery, or soft and hairy—just about any configuration.

If you are a novice, before you start to use this book take a few moments to look through the photographs and drawings and see the typical floral parts. Try to get a better feeling for what an orchid is and then go out and find some!

Which Orchid Is It?

A key is a written means for identifying an unknown species. The principles used in keying are very simple. A series of questions is asked in the form of couplets, or paired questions; the user determines a positive or negative response to each half of the pair. Be sure to read both halves of the couplet before deciding on your answer. The half of the couplet that matches your observation of the orchid directs you to the next set of couplets, and eventually to the genus and/or species name. As you progress through the key, the illustrations may help you in making your determinations.

Choose the specimen you wish to identify carefully. Look for a typical, average plant—neither the largest nor the smallest. **This key is designed so you should be able to use it without picking any of the orchids.** Only in the case of a few similar species will detailed examinations be necessary. The use of measurements has been kept to a minimum, as has the use of color. Be aware that white-flowered forms exist in many of the species and they usually occur with the typical color form.

Before you start to use the key you should always mentally note the following:

1. placement and quantity of leaves, i.e., basal vs. cauline; opposite vs. alternate; 1, 2, or more
2. placement and quantity of flowers, i.e., terminal vs. axillary; single vs. multiple
3. geographic location and habitat

Three vocabulary words that will help in your understanding of the key:

pseudobulb—the swollen storage organ at the base of the leaves, primarily on epiphytes (which do not grow in this region), occasionally on terrestrials (*Malaxis* and *Liparis*)

bract—a small, reduced leaf that usually is found on the flowering stem and/or within the inflorescence

spur/mentum—a projection at the base of the lip; it may be variously shaped from slender and pointed (spur) to short and rounded (mentum)

Other terms are illustrated within the key or can be found in the glossary on page 315.

Using the Key

If you are using the key for the first time, start with a species with which you are familiar—perhaps one of the ladies'-tresses, the genus *Spiranthes*.

Starting with couplet 1

1a lip inflated, cup, or sac shaped . . . 2
1b lip otherwise . . . 5
which takes us to couplet 5

1a

5a (green) leaves (apparently) lacking at flowering time; stem bracts may be present . . . 6
5b (green) leaves present at flowering time . . . 11
which takes us to couplet 6 (or in some species of *Spiranthes* to couplet 11)

6a plants lacking chlorophyll (no green stems or leaves evident) . . . 7
6b plants with chlorophyll (stems and leaves green) . . . 8
which takes us to couplet 8

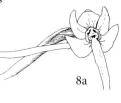
8a

8a spur or mentum present . . . 9
8b spur or mentum lacking . . . 10
which takes us to couplet 10

10a inflorescence a spike, often with many (20–60) small, tubular, white or cream-colored flowers with delicately frilled lips, usually in a loose to dense spiral . . . . . **Spiranthes**, p. 174
which then takes us the generic treatment of *Spiranthes* on page 000. Because there is more than one species in the genus, follow the same procedure with the key to the species.

This key is constructed for use in the field or with live specimens and is based on characters that are readily seen. It is not a technical key in the strictest sense, but intended simply to aid in field identification.

Keys are not difficult if you take your time, learn the vocabulary, and hone your observational skills. Like any skills, the more you use them the easier it becomes.

10a

Key to the Genera

1a lip inflated, cup or sac shaped . . . 2
1b lip otherwise . . . 5

2a

2a flowers in spikes; small, whitish in color; leaves in a basal rosette
. . . . . **Goodyera**, p. 81
2b flowers otherwise . . . 3

3a lip an inflated slipper . . . . . **Cypripedium**, p. 53
3b lip spade- or boat-shaped . . . 4

4a flowers and leaves several . . . . . **Epipactis**, p. 71
4b flower and leaf single . . . . . **Calypso**, p. 27

3a

Leaves Lacking at Flowering Time

5a (green) leaves (apparently) lacking at flowering time; stem bracts may be present . . . 6

5b (green) leaves present at flowering time . . . 11

6a plants lacking chlorophyll (no green stems or leaves evident) . . . 7

6b plants with chlorophyll (green stems or leaves evident) . . . 8

7a lip with several crest-like ridges; flower lacking a mentum . . . . . **Hexalectris**, p. 99

7b lip lacking crest-like ridges; flower with a minute mentum . . . . . **Corallorhiza**, p. 35

8a spur or mentum present . . . 9

8b spur or mentum lacking . . . 10

9a flowers brown; asymmetrical . . . . . **Tipularia**, p. 207

9b flowers green to greenish-white; symmetrical . . . . . **Piperia**, p. 125

10a inflorescence a spike, often with many (20–60) small, tubular, white or cream-colored flowers with delicately frilled lips, usually in a loose to dense spiral . . . . . **Spiranthes**, p. 174

10b inflorescence a raceme with fewer flowers (+10), not white in color . . . . . **Aplectrum**, p. 17

Leaves Present at Flowering Time

Spur or Mentum Present

11a spur or mentum present . . . 12

11b spur or mentum lacking . . . 16

12a inflorescence terminal on a leafy stem . . . 13

12b inflorescence terminal on a scape (leafless stem); flowers pink, purple, and/or white . . . . . **Galearis**, p. 77

13a lip oblong and notched at tip . . . . . ***Coeloglossum***, p. 31
13b lip otherwise; **never** notched at the tip . . . 14

14a lip entire (not divided); or with 3 teeth . . . . . ***Gymnadeniopsis***, p. 89
14b lip otherwise . . . 15

15a lip entire, divided, or fringed, but the margin **never** notched or toothed, petals erose or toothed but not fringed . . . . . ***Platanthera***, p. 128
15b petals deeply cleft; semi-aquatic . . . . . ***Habenaria***, p. 95

14a

15a

Spur or Mentum Lacking

16a pseudobulbs present, although they may be well hidden in the leaf bases . . . 17
16b pseudobulbs lacking . . . 18

17a petals and sepals similar; inflorescence a (flat-topped) corymb . . . . . ***Malaxis***, p. 121
17b petals and sepals dissimilar; petals filiform, thread-like; inflorescence a raceme . . . . . ***Liparis***, p. 109

16a

17a

17b

Leaves Basal

18a leaves essentially basal or extending up the lower 1/4 of the stem and rapidly reduced to leafy bracts . . . . . ***Spiranthes***, p. 174
18b leaves essentially cauline . . . 19

18b

Key to the Genera

Leaves Cauline

19a flowers non-resupinate, lip uppermost . . . . .
Calopogon, p. 20
19b flowers resupinate, lip lowermost . . . 20

20a leaves opposite or whorled . . . 21
20b leaves solitary or alternate (except in rare individuals) . . . 22

21a leaves 2, opposite . . . . . *Listera*, p. 115
21b leaves 5 to 7, whorled . . . . . *Isotria*, p. 103

22a leaf solitary, bracts may be present . . . 23
22b leaves several, alternate . . . . . *Triphora*, p. 211

23a inflorescence a raceme of very small, innumerable, inconspicuous green flowers . . . . . *Malaxis*, p. 121
23b inflorescence otherwise; flowers larger and showy, pink, lavender, or white; usually 1 (rarely 2–4) . . . . . *Pogonia*, p. 171

Some Important Notes About . . .

Plant Names

Example: *Platanthera leucophaea* (Nuttall) Lindley
The Latin name used for a plant consists of a genus and a species. The genus (plural, genera) is the broader group to which the plant belongs and the species is the specific plant being treated. After the two Latin names (genus and species) the name or names of the people who first described the plant are given. The example given above illustrates two different author citations. Nuttall originally described the species as *Orchis leucophaea* in 1834, therefore his name appears immediately after the genus and species. Subsequently Lindley transferred it to the genus *Platanthera* in 1835. Therefore, his name appears after Nuttall's, which has been placed in parentheses to indicate that Nuttall was the original describer but in a different genus. Two other synonyms occur for this species but they always retain Nuttall's name in parentheses: *Habenaria leucophaea* (Nuttall) Gray in 1867 and *Blephariglottis leucophaea* (Nuttall) Rydberg in 1901. In these two cases different generic concepts were used. Current nomenclature embraces the concept of *Platanthera*.

Other ranks may occur, such as subspecies, variety, and forma. Subspecies and variety usually designate a variation that has a significant difference from the species and a definite geographic range, as in *Platanthera flava* var. *herbiola*, which differs from the nominate variety in a differently shaped lip, longer floral bracts, and more northerly geographic range—hence it is a variety. Varieties and subspecies should always breed true—bearing a marked resemblance to the parent. There can always be a bit of disagreement over the rank of subspecies versus variety and which term is better used. Color variations, which occur throughout the range of the species, are best treated as forma, as in the white-flowered form of the *Calo-*

pogon oklahomensis—forma *albiflorus*. These forms can, and do, occur randomly and rarely breed true. All forma that have received names for the species occurring within the prairies and Great Plains region of North America are listed, although several have yet to be documented for this region.

Orchid enthusiasts and botanists alike will note several new forms as well as a number of "old" forms that have been resurrected. The new combinations were necessary to recognize certain taxa at the forma level. Recent research has uncovered a number of forms previously published many years ago that have been abandoned, though plants corresponding to these forms are easily found.

Common Names

Common names are never as consistent throughout the range of the plants as we would like. The most frequently used common names are used in this book, as well as some regional names.

Orchid or *orchis*? For common names either *orchid* or *orchis* may be used, but traditionally *orchis* has been used for certain genera and that has been maintained here.

Size

Average height of the plants, floral size, and number of flowers are given, with extremes in parentheses. The relative scale of each line drawing is for an average plant and is based upon the flower size.

Color

The color of the flowers as it normally occurs is given. Remember to check to see if color variants such as white-flowered forms occur. In some genera the overall color of the petals and sepals—the perianth—is given and the lip color follows.

Forms and Hybrids

The diversity of forms—randomly occurring variations such as color and habit—and hybrids are what can take hunting for native orchids to another level. After mastering the identity of many of the species, one can become challenged to search for all the variations. It would be unrealistic to think that all of these forms occur within the range of this work, but many of them do. To better understand these forms note that each is annotated with the original publication and the geographic type locality of the taxon (the geographic location from where the form was originally described). Those that have been documented within the prairies and Great Plains region of North America are often noted within the text and it is expected that several of the other forms will also be found.

Hybrids present a very different situation. Whereas forms are the result of genetic diversity within a species, hybrids rely on an outside element, the pollinator, to create the resulting hybrid as a cross between two species. In certain parental combinations hybrids are very predictable, but in other situations they are very rare. Most are obviously intermediates between the putative parents, and when both parental species are present the hybrids should be carefully sought. Few hybrids occur within the prairies and Great Plains region; they can be found primarily in the genus *Cypripedium*. The likelihood of other hybrids, especially in *Platanthera* and *Spiranthes*, which have been documented between species found in the prairies and Great Plains, is dramatically lessened because of the relative rarity of those parent species.

Flowering Periods

Flowering periods are not easy to isolate for this entire region. Latitude, exposure, and elevation play a major role in flowering times. Typically plants flower earlier in the spring in the more temperate areas and much later further north and in the mountains, whereas autumn-flowering species tend to flower earlier in the north and later southward. A good example would be *Spiranthes magnicamporum*, which flowers regularly in late August and early September in the north and well into November in the south. For stated blooming times the seasons are often noted as well as typical dates and must be adjusted for the specific area in question.

Range Maps

Both the continental range and the local range are given for each species. State and province names in brackets indicate literature reports that cannot be verified. Only species documented for the prairies and Great Plains region of a specific state or province are indicated, although the species may be found in other habitats, especially in mountainous areas. The range maps are intended to illustrate the general range of a given species and are based on verifiable specimens in herbarium records, normally housed in herbaria at any number of colleges and universities. Upon rare occasion a photograph from a reliable source, documented with date and place as a verifiable report, would be allowed. Literature reports are noted in the text and are just that: a report in any one of a number of publications that cannot be backed up by either a specimen or verifiable record. The range maps do not attempt to differentiate between extant and extirpated populations, nor do they convey how many populations are known from a given area or the size of those populations. Green dots represent widespread populations and red dots local populations. In some well-documented cases red dots bordered by black indicate historical populations. If the range of a species or variety extends beyond the boundaries of the prairies and Great Plains region of North America, as defined in this field guide, an arrow is used to indicate such.

Part 2

~

Wild Orchids of the Prairies and Great Plains Region of North America

Aplectrum

A small genus consisting of only two species, one of which is found in North America, the other in Japan. It is one of several genera found in eastern North America that have analogs in Japan.

Aplectrum hyemale (Mühlenberg ex Willdenow) Nuttall
putty-root, Adam-and-Eve

forma *pallidum* House—yellow-flowered form
Torreya 3:54. 1903 as *Aplectrum spicatum* var. *pallidum* House, type: New York

Range: Minnesota east to southern Quebec and Vermont, south to northern South Carolina and west to eastern Oklahoma

Within the prairies and Great Plains region:
Arkansas, Illinois, Iowa, Kansas, Minnesota, Missouri, Oklahoma, Wisconsin: rare to occasional

Plant: terrestrial, 18–50 cm tall

Leaves: 1; ovate, plicate, 3–8 cm wide × 10–20 cm long; appearing in the autumn and withering at flowering time in the spring

Flowers: 3–20, greenish-yellow suffused with brown and purple, the white lip three-lobed, prominently ridged, and with magenta spots or, in the forma *pallidum*, the stem and flowers yellow and the lip unspotted; individual flower size 1.5–2.0 cm

Habitat: rich, moist woodlands, bottomlands, and deciduous slopes

Flowering period: late April through early June

The **putty-root orchid**, *Aplectrum hyemale*, reaches the western limit of its range in the prairie and eastern Great Plains states. Typically a plant of rich, deciduous woodlands in the southeastern United States, **putty-root** is usually found in those woodlands that occur along some of the major river systems in the central United States. Although rare at the northern reaches of its range in southeastern Minnesota it becomes increasingly more frequent southward. The flowers are prominent, but their coloring renders them virtually invisible in the similarly hued forest surroundings. The distinctive leaf is much easier to find in late autumn, when it is nearly erect. By spring it has reclined and is usually withering when the flower spike starts to emerge. The large, pendant seed capsules also are diagnostic on the fruiting stems later in the season.

forma *pallidum*

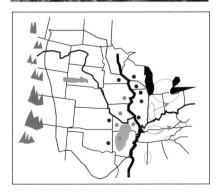

Calopogon

The genus *Calopogon* is a New World genus composed of five species, only one of which also occurs outside of the United States and Canada. Two of these species are found in the prairies and Great Plains region of North America. The non-resupinate (uppermost) lip is distinctive and easily identifies the genus.

Key to the grass-pinks, *Calopogon*

1a   flowers opening nearly simultaneously, central portion of the lip narrower than long; spring flowering plants of prairie habitats . . . . . **Oklahoma grass-pink**, *Calopogon oklahomensis*, p. 22

1b   flowers opening sequentially over a period of time, central portion of the lip wider than long; early summer flowering plants of various wetlands . . . . . **common grass-pink**, *Calopogon tuberosus*, p. 24

Calopogon oklahomensis D.H. Goldman
Oklahoma grass-pink

forma *albiflorus* P.M. Brown—white-flowered form
North America Native Orchid Journal 9: 33–34. 2003, type: Arkansas

Range: southern Minnesota east to western Indiana, south to southern Georgia and central South Carolina, and west to eastern Texas; extirpated peripherally

Within the prairies and Great Plains region:
Arkansas, Illinois, Iowa, Kansas, Louisiana, Minnesota, Missouri, Oklahoma, Texas, Wisconsin: local in remnant prairies; historical northward

Plant: terrestrial, 15–36 cm tall
Leaves: 1 or 2; lanceolate, slender, 0.5–1.5 cm wide × 7–35 cm long
Flowers: 3–7(13), non-resupinate, most of which are open simultaneously; color is highly variable, from lilac-blue to bright magenta-pink or, in the forma *albiflorus*, white, all with a golden crest on the lip; individual flower size 2.5–4.0 cm
Habitat: prairies, pine savannas, open flatwoods, and frequently mowed damp meadows
Flowering period: April throughout May (June)

The **Oklahoma grass-pink**, *Calopogon oklahomensis*, is the most recent species of grass-pink to be described from the United States. Originally thought to be restricted to the prairies of the south-central states, herbarium specimens indicate that it is much more widespread, although currently considered extirpated from much of the original range. This is one of the most variable species in coloration, from pale lilac to deep magenta and the occasional white flowered form, forma *albiflorus*. Spring in the prairies of eastern Arkansas brings a tapestry of color often with hundreds of the **Oklahoma grass-pinks** in various shades of pink.

Plants previously identified as **bearded grass-pink**, *Calopogon barbatus* (a species now know to be confined to the Deep South) or small, early-flowering **common grass-pink**, *C. tuberosus*, especially in the more northern states, occurring in suitable prairie-like habitat should be carefully examined for the possibility of **Oklahoma grass-pink**, *C. oklahomensis*. Watch for small plants with a relatively broad leaf nestled within the grasses, unlike the later-flowering **common grass-pink** of wetter areas that usually holds itself well above the surrounding vegetation. Goldman (2000; 2004) suggests that this species may have risen from ancient hybridization of *C. barbatus* and *C. tuberosus*. For more details concerning *C. barbatus* see *Wild Orchids of the Southeastern United States* (Brown and Folsom, 2004).

forma *albiflorus*

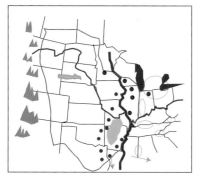

Calopogon tuberosus (Linnaeus) Britton, Sterns, & Poggenberg var. *tuberosus*

common grass-pink

forma *albiflorus* Britton—white-flowered form
Bulletin of the Torrey Botanical Club 17: 125. 1890, type: New Jersey

Range: Minnesota east to Newfoundland, south to Florida and west to Texas

Within the prairies and Great Plains region:
Manitoba; Arkansas, Illinois, Iowa, Louisiana, Minnesota, Missouri, Oklahoma, Texas, Wisconsin: occasional to frequent in all states

Plant: terrestrial, 25–115 cm tall

Leaves: 1(-3); slender, ribbed, up to 0.3–4.0 cm wide × 3–45 cm long and less than the height of the plant

Flowers: 3–17, non-resupinate, opening in slow succession; deep to pale pink or, in the forma *albiflorus*, white; all with a golden crest on the lip; individual flower size 2.5–3.5 cm

Habitat: wet meadows, pine flatwoods, damp prairies, mountain bogs, seeps, and sphagnous roadsides

Flowering period: late May through July

One of the most frequent orchids found in the eastern and central United States and Canada, the **common grass-pink**, *Calopogon tuberosus*, is a brilliant, showy plant that prefers open, wet, sandy roadsides, sphagnum bogs, and seeps. Plants flower over a very long period of time, with only a few flowers open at once. It is not unusual to find local stands of several hundred plants. Like all grass-pinks, the flowers have the lip uppermost, non-resupinate, and this feature easily separates the genus from any other with a similar morphology. Although *Calopogon tuberosus* is not a typical prairie or plains species it may occur in wetter areas or be associated with fens within the prairies. The var. *simpsonii* is found in rocky marls of the southernmost counties of Florida.

forma *albiflorus*

Calypso

The **fairy-slipper** is one of the most sought-after and delightful of all native orchids found in the Northern Hemisphere. Four varieties are known, each from its own geographic niche. Plants are found usually in dark, rich coniferous forests or, in the far north, in open tundra and barrens. The slipper-shaped flowers have a prominent and distinctive "beard" on the lip. The small white bulbs are nestled in the mosses and the single leaf, which appears in the autumn, is green throughout the winter and withers shortly after the plant flowers.

Calypso bulbosa variety *bulbosa* occurs in northern Eurasia, variety *americana* in northern North America, variety *speciosa* in Japan, and variety *occidentalis* in western North America (Wood, 1986). All differ slightly from each other and are defined geographically. The varieties *bulbosa* and *americana* are most similar as are the varieties *occidentalis* and *speciosa*.

var. *bulbosa* var. *americana* var. *occidentalis* var. *speciosa*

Calypso bulbosa (Linnaeus) Oakes var. *americana* (R.Brown) Luer

eastern fairy-slipper

> forma *albiflora* P.M. Brown—white-flowered form
> North American Native Orchid Journal 1(1): 17. 1995, type: Vermont
>
> forma *biflora* P.M. Brown—two-flowered form
> North American Native Orchid Journal 10: 35. 2004, type: Maine
>
> forma *rosea* P.M. Brown—pink-flowered form
> North American Native Orchid Journal 1(1): 17. 1995, type: Newfoundland

Range: Alaska east to Newfoundland, south in the Rocky Mountains; south to the upper Great Lakes region and northern New England

Within the prairies and Great Plains region:
South Dakota: rare in the Black Hills; local westward throughout the Rocky Mountains
Plant: terrestrial, 4–22 cm tall
Leaves: 1; strongly ribbed, ovate to cordate with a long petiole, 1.2–6.2 cm wide and 3.0–5.4(6.2) cm long
Flowers: 1, very rarely 2 in the forma *biflora*; the petals and sepals pink, erect to spreading; the lip saccate, ovate, white with dark maroon markings, or, in the forma *albiflora*, the flower white or, in the forma *rosea*, entirely pink, all with a prominent golden beard; individual flower size 2–4 cm
Habitat: calcareous woodlands often with *Thuja occidentalis*, arborvitae, and other conifers
Flowering period: late May or June to mid-July

The **eastern fairy-slipper,** *Calypso bulbosa* var. *americana*, is the undisputed showpiece of the northern woodlands and barrens, and is a widespread, although often local, orchid that deserves its ancient name of Hider-of-the-North! Whether it is growing in the deep, dark arborvitae swamps within the woodlands or nestled in the open tundra of Newfoundland, the delicate, captivating flowers demand attention. The Black Hills of South Dakota are the only place within the prairies and plains region that host the **eastern fairy-slipper** although it may be easily seen further west in the Big Horns of Wyoming and Rocky Mountains and especially in the Cypress Hills of Alberta/Saskatchewan. This is one species that always makes the wish list of new native orchid enthusiasts but, alas, is often one of the last to be seen.

forma *albiflora*

forma *rosea*

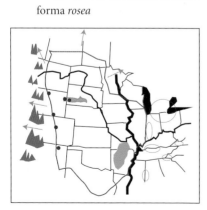

forma *biflora*

Coeloglossum

Coeloglossum is a monotypic, circumpolar genus. The plants occur in a variety of habitats in boreal, mountainous, and northern woodland areas throughout much of the Northern Hemisphere. Two varieties are known; a third, var. *interjecta*, which appears to be intermediate between the predominately Eurasian var. *viride* and the widespread North America var. *virescens*, was described by Fernald but on the basis of plants with the leaves appressed to the stem rather than wide-spreading. Recent molecular studies (Pridgeon et al., 1997; Bateman et al., 1997) have placed this *Coeloglossum* within the genus *Dactylorhiza*, but Sheviak and Catling (in *FNA*, 2002) have chosen to recognize the two genera as separate but closely related.

Coeloglossum viride (Linnaeus) Hartman var. *virescens* (Mühlenberg) Luer

long-bracted green orchis

Range: Alaska east to Newfoundland, south to Washington, New Mexico, Iowa, and North Carolina
Within the prairies and Great Plains region: Manitoba; Illinois, Iowa, Minnesota, Missouri, Nebraska, North Dakota, South Dakota, Wisconsin: rare to occasional throughout
Plant: terrestrial, 20–80 cm tall
Leaves: 3–5; 2 cm wide and up to 30 cm long passing into slender floral bracts
Flowers: 8–35; the 3–5 mm linear petals and 2–5 x 3–8 mm ovate sepals forming a hood; the 4–10 mm lip oblong and notched at the tip; flowers subtended by bracts distinctly exceeding the flowers; petals and sepals green, the lip often suffused with purple; spur minute and inconspicuous
Habitat: deciduous mesic woodlands, open coniferous forests, often along roadsides and trails
Flowering period: June through August

The **long-bracted green orchis**, *Coeloglossum viride* var. *virescens*, despite its coloration, is a conspicuous and distinctive member of the woodland orchid flora of eastern and central North America. Within the prairies and Great Plains region it may be found in wooded hedgerows, forested streamsides, and the hill country. The long, slender bracts subtending each flower give rise to the common name. Close examination of the flowers reveals the distinctive notched lip. After pollination the floral parts remain on the plant so it appears still to be in flower many weeks after flowering actually commences.

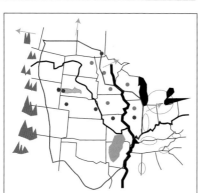

Corallorhiza

The genus *Corallorhiza* comprises 13 species found throughout North America and Hispaniola. One species, *C. trifida,* is widespread across Eurasia. The plants are entirely mycotrophic and some are thought to be saprophytes. They arise from a coralloid rhizome, hence the name. The entire genus is easily recognizable from its leafless stems, which may be variously colored, and by their small flowers. *Corallorhiza* is easily separated from *Hexalectris*, the other mycotrophic genus in eastern North America, by the lack or ridges (crests) on the lip and the presence of a mentum or small spur, although the mentum may be minute in some cases. Five species and three varieties are found within the prairies and Great Plains region.

Key to the coralroots, *Corallorhiza*

- 1a flowers cleistogamous, very small, less than 3 mm; autumn flowering . . . . . **autumn coralroot**, *Corallorhiza odontorhiza* var. *odontorhiza*, p. 40
- 1b flowers chasmogamous . . . 2
- 2a spring flowering . . . 3
- 2b summer and autumn flowering . . . 7
- 3a lip white; plain or more often with dark spots . . . 5
- 3b lip red or tan, solid or faintly striped . . . 4
- 4a flowers red to orange, tepals striped, wide-spreading . . . . . **striped coralroot**, *Corallorhiza striata* var. *striata*, p. 44
- 4b flowers dull red to tawny brown, petals and sepals not wide-spreading . . . . . **Vreeland's striped coralroot**, *Corallorhiza striata* var. *vreelandii*, p. 46
- 5a stems slender; yellow, chartreuse, or tan; early spring-flowering . . . 6
- 5b stems slender, brownish; petals and sepals indistinct, autumn flowering . . . 7
- 6a lip unlobed, spade-shaped, spotted . . . . . **Wister's coralroot**, *Corallorhiza wisteriana*, p. 50
- 6b lip with 2 lateral lobes or teeth; plain or spotted . . . . . **early coralroot**, *Corallorhiza trifida*, p. 48
- 7a stems stout; variously colored; petals and sepals distinct, late spring-summer flowering . . . 8
- 7b stems slender, brownish; plants usually under 15 cm in height; autumn-flowering . . . 9
- 8a sides of lip parallel; summer flowering . . . . . **spotted coralroot**, *Corallorhiza maculata* var. *maculata*, p. 36
- 8b sides of lip broadened, flowers open wide, late spring flowering . . . . . **western spotted coralroot**, *Corallorhiza maculata* var. *occidentalis*, p. 38
- 9a lip prominent, flowers chasmogamous . . . . . **Pringle's autumn coralroot**, *Corallorhiza odontorhiza* var. *pringlei*, p. 42
- 9b lip not prominent, often appearing undeveloped . . . . . **autumn coralroot**, *Corallorhiza odontorhiza* var. *odontorhiza*, p. 40

Corallorhiza maculata (Rafinesque) Rafinesque var. *maculata*
spotted coralroot

forma *flavida* (Peck) Farwell—yellow-stemmed form
Report (Annual) of the Regents University of the State of New York. New York State Museum 50: 126. 1897 as *Corallorhiza multiflora* var. *flavida* Peck, type: New York

forma *rubra* P.M. Brown—red-stemmed form
North American Native Orchid Journal 1(1): 8–9. 1995, type: Vermont

Range: British Columbia east to Newfoundland, south to California, Arizona, and New Mexico; Appalachian Mts. south to northern Georgia and South Carolina
Within the prairies and Great Plains region:
Manitoba; Illinois, Iowa, Minnesota, Wisconsin: very rare and local at the southern limits of its range; range extends both west and northward, where it is more common throughout
Plant: terrestrial, mycotrophic, 20–50 cm tall; stems bronzy-tan or, in the forma *flavida*, bright yellow or, in the forma *rubra*, red
Leaves: none
Flowers: 5–20; tepals typically brownish or in the forma *flavida* bright yellow or, in the forma *rubra*, red; lip white, spotted with madder purple; in the forma *flavida*, unspotted or, in the forma *rubra*, spotted with bright red; individual flowers 5.0–7.5 mm
Habitat: rich mesic and mixed forests
Flowering period: late May through July

Although the **spotted coralroot**, *Corallorhiza maculata* var. *maculata*, is the most frequently encountered species of coralroot found within North America, it is encountered less often in the northern prairies and Great Plains region. Elsewhere the variation in the stem color is evident but here, because so few plants are to be found, such variation may not be as apparent.

Corallorhiza maculata var. *occidentalis*, the **western spotted coralroot**, occurs primarily in the western and northern portions of North America and *C. maculata* var. *mexicana* in southwestern Arizona and Mexico (Coleman, 2002).

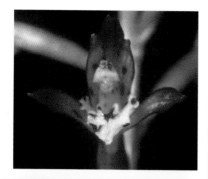

forma *flavida*

forma *rubra*

Wild Orchids of the Prairies and Great Plains Region of North America

Corallorhiza maculata (Rafinesque) Rafinesque var. *occidentalis* (Lindley) Ames

western spotted coralroot

forma *aurea* P.M. Brown—golden yellow/spotted form
> North American Native Orchid Journal 1: 195. 1995, type: Washington

forma *immaculata* (Peck) Howell—yellow spotless form
> Marin Flora, ed. 2: 363. 1970 as *Corallorhiza maculata* Rafinesque var. *immaculata* M.E. Peck, type: California

forma *intermedia* Farwell—brown-stemmed form
> Report (Annual) of the Michigan Academy of Sciences 19: 247. 1917 as *Corallorhiza maculata* (Rafinesque) Rafinesque var. *intermedia* Farwell, type: Michigan

forma *punicea* (Bartholomew) Weatherby & Adams—red-stemmed form
> Rhodora 24: 147. 1922 as *Corallorrhiza maculata* (Rafinesque) Rafinesque var. *punicea* Bartholomew, type: Michigan

Range: British Columbia east to Newfoundland, south to California, Arizona, New Mexico, Minnesota, New England, and Virginia

Within the prairies and Great Plains region:
Manitoba, Saskatchewan; Illinois, Minnesota, Nebraska, New Mexico, North Dakota, Wisconsin, South Dakota: occasional throughout, usually in open mixed woodlands

Plant: terrestrial, mycotrophic, 20–50 cm tall; stems bronzy-tan or, in the forma *aurea* and the forma *immaculata*, yellow or, in the forma *intermedia*, the stems brown or, in the forma *punicea*, the stems strikingly deep red

Leaves: none

Flowers: 5–20+; tepals typically colored bronzy-tan as the stems or, in the forma *intermedia*, the stems brown or, in the forma *punicea*, the stems strikingly deep red with the lip spotted in purple or dark red or, in the forma *immaculata*, the stems and flowers yellow to white, the lip lacking all spotting; lip 3-lobed with the middle lobe expanded, the sides obviously broadened; individual flowers 5.0–7.5 mm, the floral parts wide-spreading, mentum obscure

Habitat: rich mesic and mixed forests

Flowering period: late May through July

The common name **western spotted coralroot** for *Corallorhiza maculata* var. *occidentalis* is somewhat misleading as this variety is transcontinental. Both the nominate variety and *C. maculata* var. *occidentalis* may occur in the same woodlands and are usually well separated in flowering time except in the far north. The two varieties are easily differentiated by examining the shape of the lip. In *C. maculata* var. *maculata* the sides of the central lobe of the lip are parallel and in the var. *occidentalis* they are rounded, giving the lip a fuller form. The Black Hills of South Dakota is the best area for observing both varieties in the prairies and Great Plains region.

forma *aurea*

forma *immaculata*

forma *punicea*

Wild Orchids of the Prairies and Great Plains Region of North America

Corallorhiza odontorhiza (Willdenow) Nuttall var. *odontorhiza*

autumn coralroot

forma *flavida* Wherry—yellow-stemmed form
Journal of the Washington Academy of Science 17: 36. 1927, type: Washington, D.C.

Range: South Dakota east to Maine, south to Oklahoma and northern Florida

Within the prairies and Great Plains region:
Arkansas, Illinois, Iowa, Kansas, Louisiana, Minnesota, Missouri, Nebraska, New Mexico Oklahoma, South Dakota, Texas, Wisconsin: widespread, but never common—this plant is both rare and easily overlooked

Plant: terrestrial, mycotrophic, 5–10 cm tall; stems bronzy-green or, in the forma *flavida*, yellow

Leaves: lacking

Flowers: 5–12; cleistogamous; sepals green suffused with purple, covering the petals; lip, rarely evident in this variety, white spotted with purple or, in the forma *flavida*, unspotted; individual flower size 3–4 mm

Habitat: rich, calcareous woodlands

Flowering period: September through October

The fact that this inconspicuous little orchid is rarely found may be attributed more to its size and habit than necessarily to its rarity. The **autumn coralroot**, *Corallorhiza odontorhiza* var. *odontorhiza*, appears to be never common anywhere and is often found by accident. The short stems often flower among the fallen leaves in the autumn months and the coloration, *sans* chlorophyll, makes them even harder to see. Plants are best seen in the deciduous forests that border the many rivers in the central states and elsewhere in the hill regions.

forma *flavida*

Corallorhiza odontorhiza (Willdenow) Nuttall var. *pringlei* (Greenman) Freudenstein

Pringle's autumn coralroot

Range: Wisconsin and Ontario east to Maine, south to Iowa, Tennessee, and Georgia; Mexico, Central America
Within the prairies and Great Plains region:
Iowa, Wisconsin: rare
Plant: terrestrial, mycotrophic, 5–15 cm tall; stems bronzy-green
Leaves: lacking
Flowers: 5–12; chasmogamous; the sepals green suffused with purple, spreading and revealing the petals; lip broad, white, spotted with purple; individual flower size 5–10 mm
Habitat: rich, mesic forests and calcareous woodlands
Flowering period: September into October

Pringle's autumn coralroot, *Corallorhiza odontorhiza* var. *pringlei*, is an extremely rare variety, with its center of distribution in the lower Great Lakes. This variety has been only recently revalidated (Freudenstein, 1993, 1997), although plants with showy, chasmogamous flowers have been known for some time. In the prairies and Great Plains region it is known from a few collections from Iowa and Wisconsin. This fully open-flowered, chasmogamous form of the **autumn coralroot** is a widely scattered component of the species and may be better treated as a form.

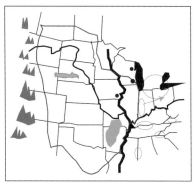

Corallorhiza striata Lindley var. *striata*

striped coralroot

   forma *eburnea* P.M. Brown—yellow/white form
   North American Native Orchid Journal 1(1): 9. 1995, type: New Mexico
   forma *fulva* Fernald—dusky tan-colored form
   Rhodora 48: 197. 1946, type: Quebec

Range: British Columbia east to Newfoundland, south to California, Texas and New York
Within the prairies and Great Plains region:
Manitoba, Saskatchewan; Nebraska, North Dakota, South Dakota: rare and local
Plant: terrestrial, mycotrophic, 20–50 cm tall; stems golden-pink with reddish-purple lines or, in the forma *eburnea*, the stems yellow or, in the forma *fulva*, dull tan
Leaves: none
Flowers: 8–35; tepals yellowish-pink with reddish-purple lines; lip white but striped purple-red with confluent lines and appearing solid or, in the forma *eburnea*, flowers yellow to white or, in the forma *fulva*, the flowers dusky-tan in color and the petals and sepals not as widespreading and the flower proportions smaller (see *Corallorhiza striata* var. *vreelandii* on p. 46); individual flowers ca. 0.75–1.5 cm
Habitat: rich mesic and mixed forests
Flowering period: late May through July

The **striped coralroot,** *Corallorhiza striata* var. *striata*, is the largest and certainly the showiest of all the coralroots in North America. A common orchid of the upper Great Lakes area, this striking species has only a few records from the northern hill country in the prairies and Great Plains region. Fernald, in 1946, described the forma *fulva* from Percé in eastern Quebec as a plant with smaller, duller tan-colored flowers—a description similar to **Vreeland's striped coralroot,** *Corallorhiza striata* var. *vreelandii*, of the western states. See the treatment of that variety on p. 46.

Magrath (in Kaul, 1986, 1989) comments on *Corallorhiza ochroleuca* Rydberg and the confusion surrounding yellow/white flowered plants in the Black Hills, noting that "Fernald's forma *fulva* should be applied to all of the yellowish flowered members of the *C. striata* complex" (Magrath, 1989). Unfortunately, it is not quite that simple and further research is needed to determine if Rydberg/Magrath's taxon *ochroleuca* is conspecific with Fernald's pale-flowered forma *fulva* and again if those plants actually represent var. *vreelandii*. Comparisons also need to be made with *C. striata* forma *eburnea*.

Plants of Rydberg's *C. ochroleuca* had not been seen for many years until May 29, 2006, when Elaine Ebbert found several dozen flowering plants of '*ochroleuca*' in an area of the Black Hills locally known as Botany Canyon. On June 10 Patti Lynch and Mary Zimmerman approached the area from another access and found many more flowering plants. No typically colored red plants of *C. striata* were seen either time in this area. This rediscovery may aid in making a future determination of the status of this taxon.

forma *eburnea*   '*ochroleuca*' Black Hills, S.Dak.

Wild Orchids of the Prairies and Great Plains Region of North America

Corallorhiza striata Lindley var. *vreelandii* (Rydberg) L.O. Williams

Vreeland's striped coralroot

forma *flavida* (Todsen & Todsen) P. M. Brown—yellow/white form
> North American Native Orchid Journal 1(1):14. 1995, type: New Mexico, *Southwestern Naturalist* 16: 122. 1971 as *Corallorhiza striata* Lindley var. *flavida* T.A. Todsen & Todsen, type: New Mexico

Range: California east to North Dakota; south to New Mexico, [Quebec]; Mexico
Within the prairies and Great Plains region:
South Dakota, Wyoming: rare and local
Plant: terrestrial, mycotrophic, 10–46 cm tall; stems pale tawny-gold with reddish-purple lines or, in the forma *flavida*, the stems yellow
Leaves: none
Flowers: 8–20; tepals dull tan with pale dusky purple lines; lip yellowish-white but striped with confluent dull purple-red lines and appearing solid in color or, in the forma *flavida*, flowers yellow to white; individual flowers ca. 0.5–0.75 cm
Habitat: rich mesic and mixed forests
Flowering period: (late April) May through late June

At first glance **Vreeland's striped coralroot**, *Corallorhiza striata* var. *vreelandii*, appears to be the poor relation of its showier cousin. Although neither the largest nor the showiest of the coralroots in North America, its delicately colored flowers are nonetheless quite beautiful. Plants are usually found in open woodlands of the Black Hills and are never common.

Fernald, in 1946, described the forma *fulva* of *C. striata* var. *striata* from Percé in eastern Quebec as a plant with smaller, duller, tan-colored flowers—a description similar to *Corallorhiza striata* var. *vreelandii* of the western states. Freudenstein (FNA, 1993, 635) states that although the forma *fulva* will key to var. *vreelandii* it really is not that closely related to the variety, but does account for distributional records for var. *vreelandii* from Quebec. See discussion of *C. ochroleuca* under *C. striata* var. *striata* for more information.

forma *flavida*

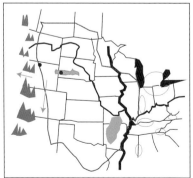

Corallorhiza trifida Chatelain

early coralroot

forma *verna* (Nuttall) P.M. Brown—yellow-stemmed/white-lipped form
> *Journal of the Academy of Natural Sciences*, Philadelphia 3: 136. 1823 as *Corallorhiza verna* Nuttall, type: Massachusetts

Range: Alaska east to Newfoundland, south to California, northern New Mexico, and in scattered localities to West Virginia

Within the prairies and Great Plains region: Manitoba, Saskatchewan; Colorado, Illinois, Minnesota, Montana, New Mexico, North Dakota, South Dakota, Wisconsin: occasional to frequent throughout

Plant: terrestrial, mycotrophic, 5–30 cm tall; stems yellow to yellow-green in the south to bronze in the far north

Leaves: none

Flowers: 8–15; tepals yellow-green to bronze, occasionally spotted with purple, wide-spreading; lip white, often spotted with purple, especially in highly colored northern individuals; mentum inconspicuous; individual flowers 0.5–1.0 cm

Habitat: rich mesic and mixed forests, open tundra and barrens

Flowering period: late May throughout much of the range to July in the north

Although smaller than many of the other species of coralroot in our area, the **early coralroot**, *Corallorhiza trifida*, with its bright greenish-yellow stems stands out among its forest companions. In the northernmost areas plants grow out in the open barrens and tundra and tend to blend in a bit more. Coloration can vary somewhat in that the plants of open exposed areas are often suffused with bronze and the floral parts with purple spots. Plants more common in the southern portion of the range have pure white lips; these were designated variety *verna* by Nuttall nearly 200 years ago, but that variation is better treated as a form.

forma *verna*

forma *verna*

Wild Orchids of the Prairies and Great Plains Region of North America

Corallorhiza wisteriana Conrad
Wister's coralroot

forma *albolabia* P.M. Brown—white-lipped form
North American Native Orchid Journal 1(1): 9–10. 1995, type: Florida

forma *cooperi* P.M. Brown—cranberry-pink–colored form
North American Native Orchid Journal 10: 22. 2004, type: Florida

forma *rubra* P.M. Brown—red-stemmed form
North American Native Orchid Journal 1(1): 62 (pl. 4). 2000, type: Florida

Range: Washington east to New Jersey, south to Arizona and Florida; Mexico
Within the prairies and Great Plains region:
Arkansas, Colorado, Illinois, Kansas, Louisiana, Missouri, Nebraska, New Mexico, Oklahoma, South Dakota, Texas: widespread and local
Plant: terrestrial, mycotrophic, 5–30 cm tall; stem brownish-yellow or, in the forma *albolabia*, yellow or, in the forma *cooperi*, cranberry-pink or, in the forma *rubra*, red
Leaves: lacking
Flowers: 5–25; sepals green, petals yellow suffused and mottled with purple; lip white, spotted with purple or, in the forma *albolabia*, yellow-stemmed, sepals and petals yellow with a pure white lip or, in the forma *cooperi*, the stems, petals, and sepals cranberry-pink, the lip white with dark pink markings or, in the forma *rubra*, red-stemmed, sepals and petals red, with flowers marked red; individual flower size 5–7 mm
Habitat: rich, often calcareous woodlands, pine flatwoods, occasionally lawns and foundation plantings
Flowering period: late March through May

Wister's coralroot, *Corallorhiza wisteriana*, with its pale brown stems and small spotted white flowers can often be seen early in the spring in open woods and even around homesites. Although individuals may be found, some locales have several thousand plants in large clustered colonies. The three color forms are exceedingly rare and known from very few sites.

forma *cooperi*

forma *albolabia*

forma *rubra*

Cypripedium

Cypripedium is a distinctive genus of about 45 species with 12 occurring in North America, north of Mexico. Although the leaf arrangement is variable, the lip, an unmistakable pouch-shaped slipper, is always diagnostic. This is often the genus first recognized by orchid enthusiasts. There are five species found within the prairies and Great Plains region of North America.

Key to the lady's-slippers, *Cypripedium*

1a leaves basal . . . . . **pink lady's-slipper**, *Cypripedium acaule*, p. 54
1b leaves cauline . . . 2

2a lip pink to purple . . . . . **showy lady's-slipper**, *Cypripedium reginae*, p. 66
2b lip white or yellow . . . 3

3a dorsal sepal arching over the lip; lip orbicular, flowers very large with the lip commonly 5.0–6.5 cm; white, ivory, or yellow . . . . . **ivory-lipped lady's-slipper**, *Cypripedium kentuckiense*, p. 58
3b lip oval to oblong . . . 4

4a dorsal sepal more or less erect; petals marked with purple, lip white, often streaked with lavender . . . . . **small white lady's-slipper**, *Cypripedium candidum*, p. 56
4b lip yellow . . . 5

5a flowers moderately large, lip ca. 3 to 5 cm long; sepals and petals unmarked to spotted, striped, or reticulately marked with reddish brown or madder; plants of a variety of habitats, usually mesic to calcareous woodlands or open sites in limestone prairies . . . . . **large yellow lady's-slipper**, *Cypripedium parviflorum* var. *pubescens*, p. 64
5b flowers small, lip 1.5–3.4 cm long . . . 6

6a lip 2.2–3.4 cm long; sepals and petals usually densely spotted with dark reddish brown appearing as a uniformly dark wash; plants of dry, deciduous, more acidic sites than *C. parviflorum* var. *pubescens* . . . . . **southern small yellow lady's-slipper**, *Cypripedium parviflorum* var. *parviflorum*, p. 60
6b sepals and petals usually suffused with dark reddish brown or madder; scent intensely sweet; plants of calcareous fens and other mesic to limey wetlands . . . . . **northern small yellow lady's-slipper**, *Cypripedium parviflorum* var. *makasin*, p. 62

Cypripedium acaule Aiton

pink lady's-slipper, moccasin flower

forma *albiflorum* Rand & Redfield—white-flowered form
   Flora of Mt. Desert Island 154. 1894, type: Maine

forma *biflorum* P.M. Brown—2–flowered form
   North American Native Orchid Journal 1: 197. 1995, type: New Hampshire

forma *lancifolium* House—narrow-leaved form
   New York State Museum Bulletin 254: 236. 1924, type: New York

Range: Northwest Territories east to Newfoundland, south to Minnesota, Mississippi, and Georgia
Within the prairies and Great Plains region:
Manitoba, Minnesota, Wisconsin: local at the western limit of the range
Plant: terrestrial, 10–55 cm tall
Leaves: 2; 5–13 cm wide × 10–30 cm long or, in the forma *lancifolium*, 3–5 cm wide; pubescent
Flowers: 1, rarely 2 in the forma *biflorum*; sepals green to reddish-brown, petals bronze; lip pale rosy-pink to deep raspberry or, in the forma *albiflorum*, white with pale green petals and sepals; individual flower size ca. 4 cm × 4 cm; lip 3–6 cm long with a longitudinal fissure
Habitat: mixed hardwood and coniferous forest; usually in highly acidic soils
Flowering period: spring, June

The **pink lady's-slipper** or **moccasin flower**, *Cypripedium acaule*, is perhaps one of the most familiar orchids found in eastern North America; it barely reaches the western limit of its range in southeastern Manitoba at the northeastern edge of the prairies and Great Plains region. Plants may occasionally be found in open woodland patches adjacent to a few prairies in Wisconsin, Minnesota, and southeastern Manitoba. Although color is variable and presents itself in just about every shade of pink, some actually tend toward peach. White-flowered plants, forma *albiflorum*, are not unusual in the northern portion of the range. Plants are notoriously difficult to transplant and re-establish, although they may persist for a few years. Resist the temptation to move plants, unless they are to be destroyed. It is far better just to admire them in their natural surroundings.

forma *lancifolium*

forma *albiflorum*

Cypripedium candidum Mühlenberg *ex* Willdenow

small white lady's-slipper

Range: Saskatchewan south to Nebraska, east to western New York, south to Missouri, Kentucky, and New Jersey; Alabama
Within the prairies and Great Plains region: Manitoba, Saskatchewan; Illinois, Iowa, Kansas, Minnesota, Missouri, Nebraska, North Dakota, South Dakota, Wisconsin: rare and local
Globally Threatened
NATIONALLY SIGNIFICANT SPECIES
Plant: terrestrial, 11–30 cm tall
Leaves: 3–5; alternate, ascending, elliptic to oblanceolate, 2–5 cm wide × 10–20 cm long
Flowers: 1–3; sepals and petals green to ochre striped with magenta, lateral sepals united; petals undulate and spiraled; lip white with delicate lavender striping beneath; individual flower size ca. 4 cm × 3 cm; lip 1.5–2.6 cm, the opening ovate at the base of the lip
Habitat: calcareous prairies and fens
Flowering period: early spring; late April in the south to early June in the north

The **small white lady's-slipper**, *Cypripedium candidum*, is one of the most imperiled of all the lady's-slippers in North America. When in flower this is the shortest in stature of the eastern lady's-slippers and one of the most highly scented. A delicate but distinct sweet odor can be detected from as far away as a meter. When the plants first emerge the flower buds are tightly enclosed within the unfolding leaves and then pop out to open while the leaves are still usually clasping. After flowering the leaves fully expand, although they still remain erect. Like the **eastern prairie fringed orchis**, *Platanthera leucophaea*, the **small white lady's-slipper** shares a dual habitat—typical mesic prairie and also calcareous fens. Further to the east both species are restricted to fens. *Cypripedium candidum* readily hybridizes with *C. parviflorum* to produce *C.* ×*andrewsii*.

Wild Orchids of the Prairies and Great Plains Region of North America   57

Cypripedium kentuckiense C.F. Reed
ivory-lipped lady's-slipper, Kentucky lady's-slipper

forma *pricei* P.M. Brown—white-flowered form
North American Native Orchid Journal 4: 45, 1998; type: Arkansas

forma *summersii* P.M. Brown—concolorous yellow-flowered form
North American Native Orchid Journal 7: 30–31, 2002; type: Arkansas

Range: northeastern Virginia; Kentucky west to Arkansas and south to eastern Texas and Georgia
Within the prairies and Great Plains region:
Arkansas, Louisiana, Oklahoma, Texas: rare and local in most areas
Plant: terrestrial, 35–98 cm tall
Leaves: 3–6; alternate, evenly spaced along the stem, broadly ovate to ovate-elliptic, 5–13 cm wide × 10–24 cm long
Flowers: 1–2 (3); sepals and petals green to yellowish-green prominently striped or spotted with dark reddish brown or, in the forma *pricei*, pale green or, in the forma *summersii*, yellow, the lateral sepals united; petals undulate and spiraled to 15 cm long; lip ivory or pale to deep yellow with delicate green spotting within or, in the forma *pricei*, white or, in the forma *summersii*, yellow and unspotted; individual flower size ca. 20 cm × 15 cm; lip 5.0–6.5 cm, the opening ovate at the base of the lip, with the overall lip compressed laterally, unlike any other of our lady's-slippers
Habitat: deciduous wooded seeps, alluvial forests, bases of slopes; occasionally in adjacent floodplains
Flowering period: late spring; mid-April to mid-May

Cypripedium kentuckiense, the **ivory-lipped** or **Kentucky lady's-slipper**, is the largest flowered of the lady's-slippers found in the United States and certainly one of the showiest. The enormous flowers are nearly twice the overall size of any of the other yellow-flowered lady's-slippers. The largest colonies are found in Arkansas; in many of the other areas the plants are few and scattered. Because these plants are so showy they have long been removed from the wild for cultivation. Both of the distinctive named color forms were found by eagle-eyed searchers: Jack Price, from northern Louisiana, and Bill Summers, author of *Orchids of Missouri*. Within the prairies and Great Plains region they are found in floodplains, deciduous woodlands, and damp hillsides, primarily in the hill country of the southern states.

forma *pricei* forma *summersii*

Cypripedium parviflorum Salisbury var. *parviflorum*

southern small yellow lady's-slipper

forma *albolabium* Magrath & Norman—white-lipped form
Sida 13: 372. 1989; type: Oklahoma

Range: Kansas east to Massachusetts, south to Arkansas and Georgia

Within the prairies and Great Plains region:
Arkansas, Illinois, Kansas, Missouri, Nebraska, Oklahoma: rare and local in the deciduous forests
Plant: terrestrial, 10–60 cm tall
Leaves: 4–5; alternate, evenly spaced along the stem, spreading; ovate to ovate-elliptic to lance-elliptic, 2.5–8.0 cm wide × 10.0–18.0 cm long; the outer surface of the lowermost sheathing bract densely pubescent when young
Flowers: 1–2(3); sepals and petals uniformly dark with dense, minute dark chestnut or reddish brown spots, often appearing a uniform color; lateral sepals united; petals undulate and spiraled to 10 cm long; lip slipper-shaped, deep rich yellow, often with scarlet markings within the lip or, in the forma *albolabium*, white; individual flower size ca. 4.5 cm × 10.0 cm; lip 2.2–3.4 cm, the opening ovate-oblong at the base of the lip
Habitat: decidedly acid, deciduous, dry to mesic, wooded slopes
Flowering period: late spring; mid-April throughout much of May

The recent separation of the two small yellow lady's-slippers has solved an often confusing question. How can small-flowered, dark-petaled, yellow lady's-slippers growing in dry, acid woods be the same as similar plants found in typical calcareous woodlands and wetlands? The answer is that they are not the same; nor are they gradations of the large yellow. The habitat, range, and delicate roselike fragrance should be more than sufficient to identify this uncommon variety. *Cypripedium parviflorum* var. *parviflorum*, the **southern small yellow lady's-slipper**, reaches the western limit of its range in the prairie regions. Unlike its cousins, the varieties *pubescens* and *makasin*, the variety *parviflorum* does not seem to venture out toward the prairies and no evidence of hybridization with *C. candidum* is evident. The curious white-lipped form, forma *albolabium*, was found in eastern Oklahoma in 1988 and may represent ancient gene flow with *C. candidum*; the latter species is not found in Oklahoma, but recently was located in southwest Missouri by Bill Summers.

forma *albolabium*

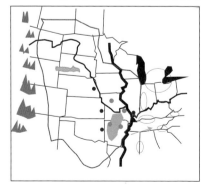

Cypripedium parviflorum Salisbury var. *makasin* (Farwell) Sheviak

northern small yellow lady's-slipper

Range: British Columbia south to northern California, east to Newfoundland, south to Illinois and Pennsylvania
Within the prairies and Great Plains region:
Manitoba, Saskatchewan; Illinois, Iowa, Minnesota, Montana, North Dakota, Wisconsin: locally scattered; usually in small populations
Plant: terrestrial, 15–35 cm tall
Leaves: 3–5; alternate, spreading; ovate to ovate-elliptic to lance-elliptic, 1.6–12.0 cm wide × 5–20 cm long; the outer surface of the lowermost sheathing bract sparsely pubescent to glabrous when young
Flowers: 1–2(3); sepals and petals suffused with a dark reddish-brown or madder, often appearing as a uniform color; lateral sepals united; petals undulate and spiraled to 10 cm long; lip ovoid, slipper-shaped, usually a deep rich yellow, with scarlet to purple markings within the lip; individual flower size ca. 2 cm × 3 cm; lip 1.5–2.9 cm long, the opening ovate-oblong at the base of the lip; intensely sweet scented
Habitat: mesic to calcareous, moist woodlands, streamsides, bogs, fens and, to a lesser degree, open heaths
Flowering period: June through July in the north

The small, richly colored and intensely fragrant flowers of the **northern small yellow lady's-slipper**, *Cypripedium parviflorum* var. *makasin*, are, in most instances, easily distinguished from those of the **large yellow lady's-slipper,** *C. parviflorum* var. *pubescens*. Confusion occurs when plants of **large yellow lady's-slipper** with rich, dark petals are found; these plants usually also have large lips. Habitat is often a help but it is important to check out all of the criteria. The intense fragrance of the **northern small yellow lady's-slipper** is a great help, and usually the flowers are decidedly smaller than those of var. *pubescens*. Hybrids with *C. candidum* are known as *C.* ×*andrewsii* nm *favillianum* and are not uncommon in the northern prairies.

Cypripedium parviflorum Salisbury var. *pubescens* (Willdenow) Knight
large yellow lady's-slipper

Range: Alaska east to Newfoundland, south to Arizona and Georgia
Within the prairies and Great Plains region:
Manitoba, Saskatchewan; Arkansas, Colorado, Illinois, Iowa, Kansas, Minnesota, Missouri, New Mexico, North Dakota, South Dakota, Wisconsin, Wyoming: locally scattered; usually in small populations
Plant: terrestrial, 15–60 cm tall
Leaves: 3–5; alternate, somewhat evenly spaced along the stem, spreading; ovate to ovate-elliptic to lance-elliptic, 2.5–12.0 cm wide × 8.0–20.0 cm long; the outer surface of the lowermost sheathing bract densely pubescent with short, silvery hairs when young
Flowers: 1–3(4); sepals and petals spotted, splotched, or marked with brown, chestnut, or reddish-brown spots, rarely appearing as a uniform color; lateral sepals united; petals undulate and spiraled to 10 cm long; lip slipper-shaped, from pale to deep rich yellow, less often with scarlet markings within the lip; individual flower size ca. 4.5 cm × 12.0 cm; lip 2.5–5.4 cm, the opening ovate-oblong at the base of the lip; scent moderate to faint reminiscent of old roses
Habitat: a wide variety of mesic to calcareous, wet to dry woodlands, streamsides, bogs, and fens
Flowering period: late May in the southern states through July in the far north

The **large yellow lady's-slipper,** *Cypripedium parviflorum* var. *pubescens,* is the classic yellow lady's-slipper so familiar to so many wildflower lovers and gardeners. Although it has declined dramatically in some areas in the past twenty-five years, it still can be found in rich forests and swamps throughout much of the mesic and calcareous woodlands of our region that border the eastern prairies. The fact that this is one of the few native orchids that can be cultivated in the garden has led to its decline in the wild. It is not all that unusual to come upon sites where, in past years, there have been many plants, only to find many holes where they have been dug. Because the plants do grow well under cultivation they can be purchased as propagated plants so there is no need to steal them from the wild! Hybrids with the **small white lady's-slipper,** *C. candidum,* are known as *C.* ×*andrewsii* nm *andrewsii* and are not uncommon in many places where *C. candidum* is found.

Wild Orchids of the Prairies and Great Plains Region of North America

Cypripedium reginae Walter
showy lady's-slipper

> forma *albolabium* Fernald & Schubert—white-flowered form
> *Rhodora* 50: 230. 1948, type: hort.

Range: Saskatchewan east to Newfoundland, south to Arkansas and southern North Carolina
Within the prairies and Great Plains region:
Manitoba, Saskatchewan; Arkansas, Illinois, Iowa, Missouri, Minnesota, North Dakota, Wisconsin: rare and local in calcareous wetlands; abundant north of the prairies and plains, especially in Minnesota
Plant: terrestrial, robust to 90 cm tall
Leaves: 3–5; alternate, somewhat evenly spaced along the stem, spreading; ovate to ovate-elliptic, densely pubescent with distinctive hairs throughout, 6.5–16.0 cm wide × 10.0–25.0 cm long
Flowers: 1–3(4); sepals and petals white; dorsal sepal arching or erect, lateral sepals united and cupped under the lip; petals oblong, wide-spreading to 4.7 cm long; lip slipper-shaped, from pale to deep rich red or maroon or, in the forma *albolabium*, white; the staminode white with contrasting yellow markings; individual flower size ca. 8 cm × 12 cm; lip globose, 2.5–5.4 cm, the opening ovate
Habitat: wet swamps, seeps, calcareous meadows, and open woodlands
Flowering period: late May in the southern states through mid-July in the north

Cypripedium reginae, the queen of the lady's-slippers, is one of the showiest and most dramatic of all of our native orchids. The plants have a preference for calcareous areas and are happy in seeps or even standing water. They also occur readily in roadside ditches. The stiff hairs on the leaves and stems are an infamous source of dermatitis and more than once have helped to identify a local poacher! The showy lady's-slipper is reaching the western limits of its range in the prairies and Great Plains states and provinces and is more scattered in its distribution, but nonetheless as dramatic. Apart from the prairies proper, plants may be abundant in the bogs and swamps of northern portions of Michigan, Wisconsin, and Minnesota.

forma *albolabium*

Wild Orchids of the Prairies and Great Plains Region of North America

Hybrids:
Three interesting hybrids have often been noted from areas where *Cypripedium parviflorum*, in variety, and *C. candidum* are sympatric.

Cypripedium ×*andrewsii* Fuller nm *andrewsii*
Andrews' hybrid lady's-slipper
(*C. candidum* × *C. parviflorum* var. *makasin*)
  Rhodora 34: 242. 1932, as *Cypripedium* ×*andrewsii*, type: Wisconsin

Cypripedium ×*andrewsii* Fuller nm *favillianum* (Curtis) Boivin
Faville's hybrid lady's-slipper
(*C. candidum* × *C. parviflorum* var. *pubescens*)
  Rhodora 34: 100. 1932, as *Cypripedium* ×*favillianum*, type: Wisconsin

Cypripedium ×*andrewsii* nm *landonii* (Garay) Boivin
Landon's hybrid lady's-slipper
(*C. candidum* × *C.* ×*andrewsii* nm *favillianum*)
  Canadian Journal of Botany 31: 660. 1953, as *Cypripedium* ×*landonii*, type: Ontario

Occasionally hybrids persist when either or both parents are no longer found in the vicinity. This is especially true with *C. candidum* as a parent.

Hybrids involving *Cypripedium candidum* and *C. parviflorum* var. *parviflorum* have not been described, although if such do exist they would also fall under the name of *C.* ×*andrewsii*.

Cypripedium ×*herae* Ewacha & Sheviak is a curious hybrid between *Cypripedium parviflorum* var. *pubescens* and *C. reginae* that has recently been described from near Winnipeg, Manitoba (*Orchids* 73(4): 296–299. 2004). Although the two parents may often grow nearby throughout much of their range, they flower together only in the far northern portion of their ranges and until now had not been documented to hybridize. The hybrid should be actively sought throughout the sympatric range of the two species. Artificial hybrids of this cross, named *Cypripedium* 'Genesis,' are available commercially.

Epipactis

Epipactis is a cosmopolitan genus of about 25 species, only one of which, *E. gigantea*, is native to North America, primarily western North America. Two Eurasian species can also be found in North America, including *E. helleborine*.

Key to the **helleborines,** *Epipactis*

1a lip 3-lobed . . . . . **stream orchid**, *Epipactis gigantea*, p. 72
1b lip not 3-lobed . . . . . **broad-leaved helleborine***, *Epipactis helleborine*, p. 74

Epipactis gigantea Douglas *ex* Hooker
stream orchid, chatterbox

forma *citrina* P.M. Brown—yellow-flowered form
North American Native Orchid Journal 7(4): 257. 2001, type: California
forma *rubrifolia* P.M. Brown—red-leaved form
North American Native Orchid Journal 1(4): 287. 1995, type: California

Range: southern British Columbia east to Montana, south to California and Arizona, New Mexico, South Dakota; Mexico
Within the prairies and Great Plains region:
New Mexico, Oklahoma, South Dakota, Texas: widely distributed and often common on lakeshores and seeps; very rare at the eastern limit of its range
Plant: terrestrial, 30–100 cm tall
Leaves: 4–12; alternate, oval-lanceolate, 2.5–7.0 cm wide × 5.5–20.0 cm long, passing into slender floral bracts, light green or, in the forma *rubrifolia*, dark red
Flowers: 5–25; sepals and petals similar, ovate-lanceolate, green to yellow striped with purple veins, the dorsal sepal and petals thrust forward, forming an arched hood over the lip, the lateral sepals wide-spreading; lip divided into two portions, the upper portion with two broad lobes, and below the constriction, the lower portion spade-shaped; yellow with purple and orange markings or, in the forma *citrina* the flower entirely yellow or, in the forma *rubrifolia*, the flowers suffused with dark red; individual flowers 1–3 cm across
Habitat: lakeshores, streamsides, and seeps
Flowering period: late June well into August

One of the most colorful members of the genus, the **stream orchid,** *Epipactis gigantea*, is the only native helleborine in North America. The slender stems hold numerous green, yellow, purple, and orange flowers throughout much of the summer months. The **stream orchid** is rare and local at the eastern limit of its range. In the Black Hills of South Dakota and eastern Wyoming it is often found around hot springs. Colonies may have several hundred stems and the only real problem in viewing these striking flowers is that you usually have to stand in the water to look back at the plants on the shoreline! Both of the interesting color forms were originally found in California by Ron Coleman (1995). The yellow-flowered forma *citrina* might occur almost anywhere the species is present. The red-leaved forma *rubrifolia* was found in a serpentine area and therefore is unlikely in the prairies and Great Plains region. It has been propagated commercially and is often available at western native plant nurseries.

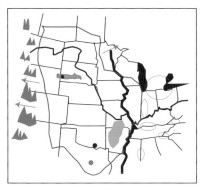

forma *citrina*

Epipactis helleborine (Linnaeus) Cranz
broad-leaved helleborine*

forma *alba* (Webster) Boivin—white-flowered form
 British Orchids, ed. 2: 21. 1898, as *Epipactis latifolia* Allioni forma *alba* Webster, type: Wales

forma *luteola* P.M. Brown—yellow-flowered form
 North American Native Orchid Journal 4: 318. 1996, type: New Hampshire

forma *monotropoides* (Mousley) Scoggin—albino form
 Canadian Field-Naturalist 41: 30. 1927, as *Amesia latifolia* A. Nelson & J.F. MacBryde forma *monotropoides* Mousley, type: Quebec

forma *variegata* (Webster) Boivin—variegated form
 British Orchids, ed. 2. 22. 1898, as *Epipactis latifolia* Allioni forma *variegata* Webster, type: England, Wales

forma *viridens* A. Gray—green-flowered form
 Botanical Gazette 4: 202. 1879, type: New York

Range: eastern North America; southeastern California; scattered in western North America; Europe
Within the prairies and Great Plains region:
Arkansas, Illinois, Minnesota, Missouri, New Mexico, Wisconsin: locally common and increasing its range
Plant: terrestrial, 10–80 cm tall, entirely white in the forma monotropoides
Leaves: 3–7; alternate, spreading; lance-elliptic, 2.5–4.0 cm wide × 10–18 cm long, green or, in the forma *variegata* with white markings or, in the forma *monotropoides*, white
Flowers: 15–50; yellow-green usually suffused with rosy-pink or, in the forma *alba*, white or, in the forma *luteola*, yellow or, in the forma *viridens*, green; individual flowers 1-3 cm across.
Habitat: highly variable, from shaded calcareous woodlands to front lawns and garden beds and even the crack in a concrete sidewalk; typically a lime-lover
Flowering period: late July through August

The widespread European **broad-leaved helleborine***, *Epipactis helleborine*, was first found in North America near Syracuse, New York, in 1878. In the ensuing century-plus it has spread throughout the region and can now be found all the way eastward to downtown Boston, Massachusetts, northward to Nova Scotia and Newfoundland and, in recent years, westward to California. To date it has been found in a few urban areas within the prairies and Great Plains region and should be carefully sought in local parks in some of the larger cities. Its fondness for calcareous areas is satisfied by the presence of concrete walkways and foundations.

forma *luteola*

forma *monotropoides*

forma *alba*   forma *viridens*

Galearis

Galearis is a small genus of two species with bractless, angled stems and two or three succulent, basal leaves. One species occurs in North America, the other in Asia. Both have showy flowers and were formerly treated in the genus *Orchis*.

Galearis spectabilis (Linnaeus) Rafinesque
showy orchis

forma *gordinierii* (House) Whiting & Catling—white-flowered form
> *Bulletin of the State Museum of New York* 243–244: 50 (1921), 1923 as *Galeorchis spectabilis* (Linnaeus) Rydberg forma *gordinierii* House, type: New York

forma *willeyi* (Seymour) P.M. Brown—pink-flowered form
> *Rhodora* 72: 48. 1970, as *Orchis spectabilis* Linnaeus forma *willeyi* Seymour, type: Vermont

Range: Minnesota east to New Brunswick, south to eastern Oklahoma and Georgia
Within the prairies and Great Plains region: Arkansas, Illinois, Iowa, Kansas, Minnesota, Missouri, Nebraska, Oklahoma, Wisconsin: local in rich woods
Plant: terrestrial, 10–15 cm tall; stem conspicuously angled
Leaves: 2(4); essentially basal, ovate to oblanceolate, 5–10 cm wide × 6–21 cm long
Flowers: 3–17; in a loose, terminal spike; floral bract conspicuous and often exceeding the flowers; petals and sepals lavender-purple, lip white; in the forma *gordinierii*, the flowers entirely white or, in the forma *willeyi*, the flowers entirely purple; individual flower size ca. 3.5 cm × 4.5 cm not including the 9–22 mm long obtuse spur
Habitat: rich mesic or calcareous woodlands, often on slopes and streamsides
Flowering period: late April in the southern states through June northward

Galearis spectabilis, the **showy orchis,** is an orchid typical of the rich mesic forest of the Appalachians; westward it is found primarily in similar woodlands in the hill and mountain areas especially of Illinois, Missouri, and Arkansas. It has a preference for rich, mesic to calcareous soils. The fleshy, almost succulent, leaves are distinctive and the plants are often the favorite rooting area for skunks, which like to feast on the tasty roots. Many a time it has been assumed that orchid thieves have dug the plants when it has been our striped friends!

forma *willeyi*

forma *gordinierii*

Goodyera

A large genus that is widespread throughout the world, *Goodyera* is known for its beautifully marked and reticulated leaves that often earn the group the name of "jewel orchids." The degree of leaf marking varies greatly even within a species. In Canada and the United States we have four species, all primarily northern or higher elevation in distribution. Three species occur within the prairies and Great Plains region.

Key to the **rattlesnake orchises**, *Goodyera*

1a flowers in a dense spike . . . . . **downy rattlesnake orchis**, *Goodyera pubescens*, p. 84
1b flowers in a lax, often 1-sided spike . . . 2

2a flowers white; rosettes usually not more than a few, 2.5–4.0, cm across . . . . . **lesser rattlesnake orchis**, *Goodyera repens*, p. 86
2b flowers tan to white; rosettes 5–10 cm across . . . . . **giant rattlesnake orchis**, *Goodyera oblongifolia*, p. 82

Goodyera oblongifolia Rafinesque
giant rattlesnake orchis

forma *reticulata* (Boivin) P.M. Brown—reticulated leaved form
> *Canadian Field-Naturalist* 65: 20. 1951 as *Goodyera oblongifolia* Rafinesque var. *reticulata* B. Boivin, type: British Columbia

Range: southeastern Alaska east to Newfoundland, south to Maine, California, and New Mexico; Mexico
Within the prairies and Great Plains region:
Colorado, Nebraska, New Mexico, North Dakota, South Dakota, Wisconsin, Wyoming: rare and local in the east, becoming more frequent westward
Plant: terrestrial, to 45 cm tall
Leaves: 3–7; in a basal rosette, dull bluish-green with a white central stripe or, in the forma *reticulata*, with handsome white reticulations; lanceolate, 2–4 cm wide × 5–11 cm long
Flowers: 10–30+; in a tall, loosely flowered one-sided (secund) terminal spike; white with coppery-green overlays; dorsal sepal and petals forming a hood over the pointed lip, lateral sepals reflexed; individual flower size 4 mm × 5 mm
Habitat: dry, mixed and deciduous woodlands
Flowering period: August

The **giant rattlesnake orchis**, *Goodyera oblongifolia*, is the largest of all of the rattlesnake orchises in North America in both rosette size and height. Unfortunately it is also one of the less showy of the genus. The broad, dull rosettes occasionally make colonies but more usually occur singly. The tall, few-flowered spike carries its off-white flowers in a loose, one-sided raceme.

In the reticulated leaf form, forma *reticulata*, the leaves take on a handsome appearance with beautiful, delicate reticulations. This variation can, and does, occur randomly throughout the range of the species, although it is more frequently encountered in western North America. The **giant rattlesnake orchis** is one of the northern/western disjunct features of the Black Hills of South Dakota and eastern Wyoming.

forma *reticulata*

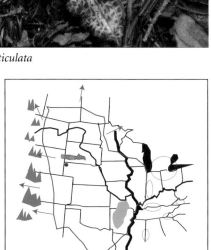

Goodyera pubescens (Willdenow) R. Brown

downy rattlesnake orchis

Range: Ontario east to Nova Scotia, south to Arkansas and Florida
Within the prairies and Great Plains region:
Arkansas, Illinois, Iowa, Minnesota, Missouri, Oklahoma, Wisconsin: very rare and local at the western limit of its range
Plant: terrestrial, 20–50 cm tall
Leaves: 4–8; in a basal rosette, bluish-green with white reticulations on the veins, broadly lanceolate, 2–4 cm wide × 4–10 cm long; evergreen
Flowers: 20–50+; in a densely flowered terminal spike; white, copiously pubescent; petals and sepals somewhat similar, the upper ones forming a hood over the spreading sepals and saccate lip; individual flower size ca. 3 mm × 4 mm
Habitat: mixed and deciduous woodlands
Flowering period: August

The **downy rattlesnake orchis**, *Goodyera pubescens*, is typically a species of the eastern temperate forest that is reaching the western limit of its range in a few of the prairies and Great Plains region states. It has the most handsomely marked foliage of any of our native orchids with the added feature of being evergreen. Large patches are often formed; when in flower the numerous snow-white blooms atop the slender spikes make it the showiest of all of the rattlesnake orchises in North America. The entire inflorescence is copiously pubescent and the neat, little rounded buds form a fanciful appearance similar to that of the tail of a rattlesnake!

Wild Orchids of the Prairies and Great Plains Region of North America 85

Goodyera repens (Linnaeus) R. Brown
lesser rattlesnake orchis

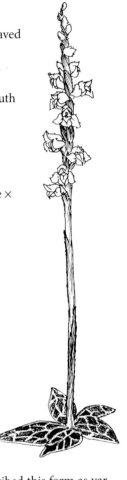

forma *ophioides* (Fernald) P.M. Brown—white-veined leaved form

Rhodora 1: 6. 1899 as *Goodyera repens* var. *ophioides* Fernald, type: northeastern United States and Canada
Range: Alaska east to Newfoundland, south to Wyoming, south in the Rocky and Appalachian Mountains; northern Eurasia
Within the prairies and Great Plains region:
South Dakota: local, primarily in the forma *ophioides*
Plant: terrestrial, 5–23 cm tall
Leaves: 3–6; in a basal rosette, dark green or, in the forma *ophioides*, marked with silver veining, ovate, 0.5–2.0 cm wide × 1.0–4.2 cm long
Flowers: 10–20; in a loosely flowered, one-sided, terminal raceme; white, pubescent; dorsal sepal and petals forming a hood over the rounded lip, lateral sepals often reflexed; individual flower size 2 mm × 3(4) mm
Habitat: mixed and deciduous woodlands
Flowering period: late July through August

The **lesser rattlesnake orchis,** *Goodyera repens,* is the tiniest of the rattlesnake orchises, and is the only North American species found in Eurasia as well. The small rosettes of the nominate variety are nearly plain with little or no contrasting veining. This is the leaf coloring seen throughout most of the range in Europe and Asia but rarely in northernmost North America. The forma *ophioides,* with beautiful silver veining on the leaves, is the form most frequently seen through most of North America. Plants can be quite variable in the degree of veining. Fernald described this form as var. *ophioides* but as it passes into var. *repens* northward it is best treated as a form.

forma *repens*

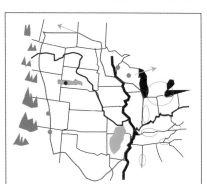

forma *ophioides*

Wild Orchids of the Prairies and Great Plains Region of North America

Gymnadeniopsis

Gymnadeniopsis was formerly placed within the genera *Habenaria* and *Platanthera* (Rydberg, 1901). The three species that Rydberg used to create the genus *Gymnadeniopsis* have recently been reassessed (Sheviak in FNA, 2002; Brown, 2003). Several differences are present that render them distinctive, including the presence of tubers on the roots as well as small tubercles on the column. The three members of this genus have occasionally been known as the frog orchids; there are two species found in the prairies and Great Plains region.

Key to the frog orchids, *Gymnadeniopsis*

1a flowers white; non-resupinate (lip uppermost) . . . . . **snowy orchis**, *Gymnadeniopsis nivea*, p. 92
1b flowers green to straw-colored to nearly white; spur swollen at the tip . . . . . **little club-spur orchis**, *Gymnadeniopsis clavellata*, p. 90

Gymnadeniopsis clavellata (Michaux) Rydberg var. *clavellata*
little club-spur orchis

forma *slaughteri* (P.M. Brown) P.M. Brown—white-flowered form
> North American Native Orchid Journal 1(3): 200. 1995, as *Platanthera clavellata* var. *clavellata* forma *slaughteri* P.M. Brown, type: Arkansas

forma *wrightii* (Olive) P.M. Brown—spurless form
> Bulletin of the Torrey Botanical Club 78(4): 289–291. 1951, as *Habenaria clavellata* (Michaux) Sprengel var. *wrightii* Olive, type: North Carolina

Range: Wisconsin east to Maine, south to Texas and Georgia
Within the prairies and Great Plains region:
Arkansas, Illinois, Iowa, Louisiana, Missouri, Oklahoma, Texas, Wisconsin: occasional to rare
Plant: terrestrial, 15–35 cm tall
Leaves: 2; cauline, ovate-lanceolate, 1–2 cm wide × 5–15 cm long, passing upward into bracts
Flowers: 5–15; arranged in a loose terminal raceme, flowers usually twisted to one side; sepals ovate, petals linear, enclosed within the sepals and forming a hood; lip oblong, the apex obscurely 3-lobed; perianth yellow-green or, in the forma *slaughteri*, white; individual flower size 0.5 cm, not including the 1 cm spur, the small tip swollen (clavate) or, in the forma *wrightii*, the spur absent and the lip untoothed
Habitat: damp woods, streamsides, open wet ditches
Flowering period: June through August

The **little club-spur orchis**, *Gymnadeniopsis clavellata*, is a northeastern species that becomes increasingly rare as its range progresses southward and westward. The small, pale greenish flowers of the **little club-spur orchis** are very different from any other orchid we have, and they hold themselves at curious angles on the stem. The distinctive spur, with its swollen tip, is what gives this plant its common name. Plants of the nominate variety are found most often in wooded swamps and occasionally in open fens and mesic prairies. *Gymnadeniopsis clavellata* var. *ophioglossoides* is usually found in the northern and higher elevation areas of northeastern North America.

forma *slaughteri*

Gymnadeniopsis nivea (Nuttall) Rydberg

snowy orchis

Range: southern New Jersey south to Florida, and west to Texas

Within the prairies and Great Plains region:
Arkansas, Louisiana, Texas: rare and local; presumed extirpated in Arkansas
Plant: terrestrial, 20–60 cm tall
Leaves: 2–3; cauline, lanceolate, keeled, conduplicate, 1–3 cm wide × 5–25 cm long
Flowers: 20–50; non-resupinate, arranged in a densely flowered terminal raceme; sepals and petals oblong-ovate; lip uppermost, linear-elliptic and bent backward midway; perianth stark, icy white; individual flower size 8–10 mm, not including the slender 1.5 cm spur
Habitat: open wet meadows, prairies, seeps, damp pine flatwoods
Flowering period: late May through July

The **snowy orchis**, *Gymnadeniopsis nivea*, is a typical southeastern species and occurs in scattered locales throughout the Coastal Plain, often in prairie-like habitats and savannas such as those of western Louisiana, eastern Texas and (formerly) southern Arkansas. Recently burned pine flatwoods are also an excellent habitat to look for this orchid. In addition, it has the added feature of being deliciously fragrant. The uppermost lip is unique among the species of *Gymnadeniopsis*, *Habenaria*, and *Platanthera* found in the southern prairies and Great Plains.

Wild Orchids of the Prairies and Great Plains Region of North America   93

Habenaria

The genus *Habenaria* is pantropical and subtropical and consists of about 600 species. It reaches the northern limit of its range in North America in the southeastern United States. In its broadest sense, it has often included those species that are currently found in *Platanthera, Coeloglossum, Gymnadeniopsis,* and *Pseudorchis*. As treated here, in the narrow sense, the genus contains five species, three of which are found within the southeastern United States and the other two confined to central and southern Florida. Only one species, *H. repens*, is found in the prairies and Great Plains region.

Habenaria repens Nuttall
water-spider orchis

Range: North Carolina south to Florida, west to southeastern Arkansas and Texas; Mexico, West Indies, Central America
Within the prairies and Great Plains region:
Arkansas, Louisiana, Oklahoma, Texas: rare to local
Plant: terrestrial or aquatic, up to 50 cm tall
Leaves: 3–8; yellow-green, linear-lanceolate 1.0–2.5 cm wide × 3–20 cm long; rapidly reduced in size and passing to bracts within the inflorescence
Flowers: 10–50; in a densely flowered terminal raceme; sepals light green, ovate to oblong; petals greenish-white, with 2 divisions; lip with 3 divisions, the central being shorter than the lateral divisions; spur slender, ca. 1.3 cm long; individual flower size ca. 2 cm × 2 cm, not including the spur
Habitat: open shorelines, seasonally wet grasslands, stagnant pools
Flowering period: primarily May through October but occasionally throughout the year

The **water-spider orchis,** *Habenaria repens,* is one of the few truly aquatic orchids found in North America. Masses of several hundred floating plants can often be found, frequently colonizing wet roadside ditches and canals. It is the most far-ranging of the *Habenaria* species in the United States and is widespread throughout the Southeast, reaching the northwestern limit of its range in the prairies and Great Plains region within southeastern Oklahoma.

Hexalectris

A genus of seven species found primarily in the southern United States and Mexico, *Hexalectris* is similar in appearance to the other mycotrophic genus, *Corallorhiza*. Although not closely related, they both have colorful flowers terminating a leafless stem that lacks all chlorophyll. The crested flowers on *Hexalectris* are much larger than those of *Corallorhiza*, and intricately more beautiful. The only *Hexalectris* species in the prairies and Great Plains region, the crested coralroot, *Hexalectris spicata*, is also the only species ranging east of Texas from the southwestern United States and adjacent Mexico.

Hexalectris spicata (Walter) Barnhardt var. *spicata*
crested coralroot

> forma *albolabia* P.M. Brown—white-lipped form
> *North American Native Orchid Journal* 1(1): 10. 1995, type: Florida
> forma *lutea* P.M. Brown—yellow-flowered form
> *North American Native Orchid Journal* 10: 23. 2004, type: Florida
> forma *wilderi* P.M. Brown—albino form
> *North American Native Orchid Journal* 10: 23. 2004, type, Florida

Range: Arizona, Missouri; southern Illinois east to Maryland, south to Florida and west to Texas; Mexico
Within the prairies and Great Plains region:
Arkansas, Illinois, Kansas, Louisiana, Missouri, New Mexico, Oklahoma, Texas: widespread, although never common
Plant: terrestrial, mycotrophic, 10–80 cm tall; stems yellow-brown to deep purple or, in the forma *lutea*, the stems yellow or, in the forma *wilderi*, the stems white
Leaves: lacking
Flowers: 5–25; sepals and petals brown-yellow with purple striations; lip pale yellow with purple stripes (crests), 3-lobed, the lateral lobes incurved or, in the forma *albolabia*, the lip pure white with pale yellow striations and the petals and sepals mahogany or, in the forma *lutea*, the flowers yellow or, in the forma *wilderi*, the flowers white; individual flower size 2.5–4.0 cm
Habitat: dry, open hardwood forest especially under live oak
Flowering period: June in the south to August in the more northerly states

The **crested coralroot,** *Hexalectris spicata*, is by far the handsomest of all the mycotrophic orchids (crested coralroots and coralroots), and one of the most beautiful of the summer-flowering orchids in the south-central and southeastern United States. One of several species of entirely mycotrophic orchids found in North America, the plants lack all traces of chlorophyll and therefore often blend in with their surroundings. From a distance they may appear to be just dead sticks, but upon closer examination the striking and colorful flowers reveal an intricate pattern of crests upon the lips. Plants have a preference for live oak woodlands and the number of plants, depending on the rainfall in a given season, may vary greatly from year to year. *Hexalectris spicata* var. *arizonica* occurs from central Texas westward.

forma *lutea*

Wild Orchids of the Prairies and Great Plains Region of North America

Isotria

The genus *Isotria* consists of only two species, both of which are found in the eastern half of the United States and adjacent Canada. They are related to the genera *Pogonia* and, more distantly, *Triphora*, and early in their history were placed in the genus *Pogonia*. We have both species in the prairies and Great Plains region.

Key to the whorled pogonias, *Isotria*

    1a   sepals greenish-yellow; one and one-half times as long as the petals or shorter than the petals . . . . . **small whorled pogonia**, *Isotria medeoloides*, p. 104

    1b   sepals purple; two or more times as long as the petals . . . . . **large whorled pogonia**, *Isotria verticillata*, p. 106

Isotria medeoloides (Pursh) Rafinesque

small whorled pogonia

Range: Michigan east to Maine, south to Missouri and South Carolina
Within the prairies and Great Plains region:
Illinois, Missouri: very rare in all states in which it occurs, often no more than a few plants
FEDERALLY LISTED AS THREATENED
Plant: terrestrial; 8–12 cm tall in flower; mature plants expand up to 15 cm tall after flowering
Leaves: 5 or 6; in a whorl at the top of the stem, up to 1 cm wide × 5 cm long
Flowers: 1 or 2; sepals and petals greenish-yellow, wide spreading; lip white; individual flowers ca. 2–3 cm across
Habitat: various wooded habitats favoring beech, mixed pines, etc., often near seasonal runoffs
Flowering period: April through May, usually before the trees fully expand their leaves

Isotria medeoloides, the **small whorled pogonia,** was one of the first orchids listed by the federal government under the Endangered Species Act. In the prairies and Great Plains region it is known only from Missouri and Illinois. The **small whorled pogonia** has a preference for wooded slopes along riverbanks and areas with seasonal runoffs. The plants nearly double in size after flowering and in the autumn they turn a bright yellow. This coloration coupled with the usually present distinctive fruit make discovery more likely at that time of year than when the diminutive plants are in flower. The species is very rare and often known from a single station, or even from a single plant in most of the states in which it occurs in North America. Only in the Northeast—New Hampshire and western Maine—can the **small whorled pogonia** be locally abundant. Both of these northeastern states have sites in excess of several thousand plants. This is an excellent example of a species, long thought to be one of the very rarest in North America, turning up in new sites because more people, both professional and amateur, have been searching for it. Extant populations of **small whorled pogonia** should be actively sought in Illinois, Missouri, and Arkansas.

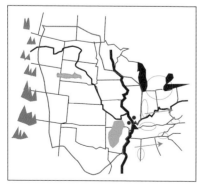

Wild Orchids of the Prairies and Great Plains Region of North America   105

Isotria verticillata (Mühlenberg *ex* Willdenow) Rafinesque

large whorled pogonia

Range: Michigan east to Maine, south to Texas and Florida
Within the prairies and Great Plains region:
Arkansas, Illinois, Louisiana, Missouri, Oklahoma,
Texas: locally common in many areas
Plant: terrestrial; to 20 cm tall in flower, mature plants expanded up to 40 cm tall
Leaves: 5 or 6; in a whorl at the top of the stem, up to 3 cm wide × 9 cm long
Flowers: 1, rarely 2; sepals purplish, wide-spreading, slender and spidery; petals pale yellow, ovate, arched over the column; lip white, edged in purple and with a fleshy, yellow central ridge; individual flowers ca. 10 cm across
Habitat: deciduous forests
Flowering period: late April through much of May, usually before the trees fully expand their leaves

The **large whorled pogonia,** *Isotria verticillata*, will often form colonies with considerable numbers of stems because the plants produce stolons. Seed set from the flowers is minimal and few fruits are usually found in these large colonies. The fanciful flowers of *I. verticillata* are unmistakable and the plants have an unusual habit (as does the only other species in the genus, *I. medeoloides*) of nearly doubling in both leaf surface and height after flowering.

Wild Orchids of the Prairies and Great Plains Region of North America

Liparis

Liparis is a cosmopolitan genus of more than 200 species and occurs in the widest variety of habitats throughout the world. All members of the genus are terrestrial or semi-epiphytic and have swollen leaf bases that form pseudobulb-like structures. These features are not unlike those of the genus *Malaxis*; they are more evident and usually aboveground in the subtropical and tropical species, whereas in the temperate and more northerly species the structure is nestled within the ground. Three species occur in the United States and Canada, but only two within the prairies and Great Plains region.

Key to the greater twayblades, *Liparis*

- 1a flowers chocolate-purple; plants of rich woodlands . . . . . **lily-leaved twayblade**, *Liparis liliifolia*, p. 110
- 1b flowers yellow-green; plants usually of moist areas, borrow pits, grassy fens, etc . . . . . **Loesel's twayblade**, *Liparis loeselii*, p. 112

Liparis liliifolia (Linnaeus) Richard *ex* Lindley
lily-leaved twayblade

forma *viridiflora* Wadmond—green-flowered form
Rhodora 34: 18. 1932, type: Wisconsin

Range: Minnesota and Ontario east to New Hampshire, south to Oklahoma and Georgia
Within the prairies and Great Plains region: Arkansas, Illinois, Iowa, Minnesota, Missouri, Oklahoma, Wisconsin: scattered populations that can be locally abundant
Plant: terrestrial, 9–30 cm tall
Leaves: 2; basal, green, strongly keeled, ovate 4 cm wide × 6–8 cm long
Flowers: 5–78; in a terminal raceme; sepals purple, ovate; petals and sepals green, slender and thread-like; lip chocolate-purple, broadly ovate or, in the forma *viridiflora*, flowers entirely green; individual flower size 1.2–2.4 cm
Habitat: rich, damp woodlands, mesic forest, shaded banks and roadsides, often occurring in calcareous soils
Flowering period: late spring; May through June

The **lily-leaved twayblade**, *Liparis liliifolia*, is a broad, handsome orchid of rich mesic forests and streamsides. It is one of the prime components of the southern Appalachian woodlands and continues to be found westward to the valleys of the great rivers within the prairie region. It is widespread in its distribution, often found growing with several other orchid species including the **putty-root**, *Aplectrum hyemale*, and **showy orchis**, *Galearis spectabilis*.

forma *viridiflora*

Liparis loeselii (Linnaeus) Richard
Loesel's twayblade, fen orchis

Range: British Columbia east to Nova Scotia, south to Arkansas, Mississippi, and the southern Appalachian Mountains; Europe
Within the prairies and Great Plains region:
Manitoba, Saskatchewan; Arkansas, Illinois, Iowa, Kansas, Minnesota, Missouri, Nebraska, South Dakota, Wisconsin: a northern species that reaches the southern limit of its range in the central prairies and Great Plains states
Plant: terrestrial, 4–20 cm tall
Leaves: 2; basal, pale green, strongly keeled; oblanceolate, 2–3 cm wide × 4–6 cm long
Flowers: 5–15; in a terminal raceme; sepals and petals slender and thread-like; lip, broadly ovate; watery-green; individual flower size 0.5–1.0 cm
Habitat: damp gravels; bogs; ditches, seepages, shaded banks and roadsides often in calcareous soils
Flowering period: late spring; May through June

The **fen orchis**, *Liparis loeselii*, is one of the few species that the eastern United States shares with northern Europe. And as rare as it is in Europe, it can be common in northeastern North America and is broadly distributed in many of the prairie and plains states. The flowers of the **fen orchis** can be easily overlooked because of their translucent coloring. In the prairies and plains region it occurs in widely scattered colonies and is most often seen in prairie fens. Should it occur with *L. liliifolia*, the **lily-leaved twayblade**, it may produce the very rare hybrid *Liparis ×jonesii*, **Jones' hybrid twayblade**.

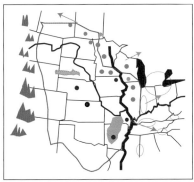

Hybrid:
Liparis ×jonesii S. Bentley
Jones' hybrid twayblade
(*L. liliifolia* × *L. loeselii*)
> *Native Orchids of the Southern Appalachian Mountains*, pp. 138–39. 2000, type: North Carolina

Although both parent species occur in many places in the eastern and central United States, this unusual hybrid has been documented only from the type locality along the Blue Ridge Parkway of southwestern North Carolina.

Listera

The genus *Listera* comprises 25 species that occur in the cooler climes of both the Northern and Southern Hemispheres. Eight species in the genus grow in the United States and Canada, one of which, *Listera ovata*, is a very common species in Europe that has become naturalized in southern Ontario. We have only two species in the prairies and Great Plains region. Recent molecular research has influenced some authors to nest *Listera* within the genus *Neottia* (Pridgeon et al., 2005).

Key to the lesser twayblades, *Listera*

1a lip cleft to more than 1/3 its length, segments narrow . . . . . **southern twayblade**, *Listera australis*, p. 116
1b lip broadened at tip and cleft to less than 1/3 its length, segments broad . . . . . **broad-lipped twayblade**, *Listera convallarioides*, p. 118

Listera australis Lindley
southern twayblade

forma *scottii* P.M. Brown—many-leaved form
North American Native Orchid Journal 6(1): 63–64. 2000, type: Florida

forma *trifolia* P.M. Brown—3-leaved form
North American Native Orchid Journal 1(1): 11. 1995, type: Vermont

forma *viridis* P.M. Brown—green-flowered form
North American Native Orchid Journal 6(1): 63–64. 2000, type: Florida

Range: Quebec to Nova Scotia, and south to Florida and Texas
Within the prairies and Great Plains region:
Arkansas, Louisiana, Oklahoma, Texas: local throughout
Plant: terrestrial in damp soils, up to 35 cm tall
Leaves: 2; opposite, midway on the stem, green, usually flushed with red, ovate 2.0 cm wide × 3.5 cm long or, in the forma *scottii*, leaves several, scattered along the stem or, in the forma *trifolia*, leaves 3, in a whorl
Flowers: 5–40; in a terminal raceme; sepals purple, ovate; petals purple, narrowly spatulate, recurved; lip purple, linear, split into 2 slender filaments or, in the forma *viridis*, flowers entirely lime-green; individual flower size 6–10 mm
Habitat: rich, damp woodlands, often in sphagnum moss
Flowering period: late January to March

A spring ephemeral, the **southern twayblade**, *Listera australis*, appears quickly in late February and March. Plants have a preference for damp, often seasonally flooded, deciduous woodlands. In many sites, it prefers the presence of sphagnum moss. Although most populations consist of less than a dozen plants, occasional sites may have several thousand individuals and contain an amazing degree of variation with all described forma present. The plants at many of these larger sites tend to be more robust than elsewhere, flowering over a very long period of time and forming dense, lush clumps. The species sets seed and senesces very quickly so a month after blooming there usually is no sign of the plants until the next flowering season.

forma *viridis*

forma *scottii*

Listera convallarioides (Swartz) Nuttall
broad-lipped twayblade

forma *trifolia* P.M. Brown—3-leaved form
North American Native Orchid Journal 1(1): 11. 1995, type: Newfoundland

Range: southwestern Alaska; British Columbia east to Newfoundland, south to California and Wyoming, east to northern Michigan and Maine
Within the prairies and Great Plains region: South Dakota, Wyoming: rare and local; confined to the Black Hills region within the prairies and plains
Plant: terrestrial; 10–30 cm tall
Leaves: 2; opposite, midway on the stem or, in the forma *trifolia*, 3 in a whorl, green, ovate-oblong, 1–3 cm wide × 2–6 cm long
Flowers: 5–15; in a terminal raceme; sepals and petals watery-green, reflexed; lip pale yellow-green, oblong, broadening to a deep cleft at the apex; pedicles and ovary pubescent; individual flower size 0.6–1.5 cm
Habitat: damp to wet, cold, mossy woodlands, thickets, and river shores
Flowering period: July

The **broad-lipped twayblade**, *Listera convallarioides*, is the largest-flowered of the native twayblades (*Listera* spp.) that we have in North America. The plants are almost always colonial and may form large patches, usually in open damp woods, in mossy glades, and on isolated little islets in flowing streams. *Listera convallarioides* is often mistakenly called the broad-leaved twayblade, but the correct common name is the **broad-lipped twayblade**, referring to the decidedly broadened apex of the lip that is unlike that of any other *Listera* in North America. Within the prairies and Great Plains region *L. convallarioides*, along with the **eastern fairy-slipper**, *Calypso bulbosa* var. *americana*, and **slender bog orchis**, *Platanthera stricta*, are northern and western species confined to disjunct locales in the mountains of the Black Hills of South Dakota and eastern Wyoming.

Malaxis

The genus *Malaxis* is cosmopolitan, with about 300 species. Eleven species are found in Canada and the United States, with only two in our region. All *Malaxis* species have a pseudobulbous stem, which is more evident in the subtropical and tropical species. In the temperate species it appears more corm-like. The genus possesses some of the smallest flowers in the Orchidaceae, many not more than a few millimeters in any dimension.

Key to the adder's-mouths, *Malaxis*

1a inflorescence a flat-topped cluster; flowers green . . . . . **green adder's-mouth**, *Malaxia unifolia*, p. 122
1b inflorescence a slender spike; flowers greenish-white . . . . . **white adder's-mouth**, *Malaxis brachypoda*, p. 120

Malaxis brachypoda (Gray) Fernald

white adder's-mouth

Range: Minnesota east to southern Quebec and Vermont, south to northern South Carolina and west to eastern Oklahoma
Within the prairies and Great Plains region: Illinois
Plant: terrestrial; 8–25+ cm tall, stem swollen at the base into a (pseudo)bulb
Leaves: 1; ovate, keeled, to 6 cm wide × 9 cm long, midway on the stem; green
Flowers: 5–80+; arranged in a slender spike; sepals oblanceolate, yellow-green to whitish-green; petals linear and positioned behind the flower; lip, white, broadly triangular; individual flower size ca. 1.5–3.0 mm
Habitat: fens and sphagnum bogs
Flowering period: late June

Malaxis brachypoda (syn. *Malaxis monophyllos* (Linnaeus) Swartz var. *brachypoda* (Gray) Morris & Eames) is reported for Illinois by Sheviak (1974) based upon a specimen labeled from Elgin Fen, Kane County, without a collection date. Some confusion surrounds this record, and Sheviak goes into detail concerning other possible localities. The **white adder's-mouth** is not found elsewhere in the prairies and Great Plains region of North America, although it does occur in non-prairie portions of Wisconsin, Minnesota, and Manitoba. It is not included in Winterringer (1967), Case (1987), or in the Illinois list of rare plants species (see page 248) and has not received full treatment within this book as a species found within the prairies and Great Plains, although the Illinois record must be noted. See *Wild Orchids of the Canadian Maritimes and North Great Lakes Regions* (Brown and Folsom, 2006) for a full treatment of this species and *Wild Orchids of the Pacific Northwest and Canadian Rockies* (Brown and Folsom, 2006) for details concerning the synonyms.

Malaxis unifolia Michaux
green adder's-mouth

forma *bifolia* Mousley—2-leaved form
Orchid Review 35: 163. 1927, type: Quebec

forma *variegata* Mousley—variegated-leaf form
Orchid Review 35: 164. 1927, type: Quebec

Range: Manitoba east to Newfoundland south to Texas and Florida; Mexico

Within the prairies and Great Plains region: Manitoba; Arkansas, Illinois, Iowa, Kansas, Louisiana, Missouri, Oklahoma, Texas, Wisconsin: local to occasional but often overlooked

Plant: terrestrial; 8–25+ cm tall, stem swollen at the base into a (pseudo)bulb

Leaves: 1 or, in the forma *bifolia*, 2; ovate, keeled, to 6 cm wide × 9 cm long, midway on the stem; green or, in the forma *variegata*, with white markings

Flowers: 5–80+; arranged in a compact raceme, elongating as flowering progresses; sepals oblanceolate, green; petals linear and positioned behind the flower; lip green, broadly ovate to cordate, with extended auricles at the base and bidentate at the summit; individual flower size 2–4 mm

Habitat: damp woodlands, moist open barrens, mossy glades, fens, and sphagnum bogs

Flowering period: late June through much of August

Often considered one of the most widespread and common orchids in eastern North America, the **green adder's-mouth**, *Malaxis unifolia*, may be a real challenge to find. Plants vary greatly in size and their natural camouflage blends in with much of the surrounding vegetation. Only when growing in open mossy barrens do they really stand out. The **green adder's-mouth** is widely scattered within the prairies and Great Plains region and easily overlooked. Large plants are not uncommon and they, like most members of the genus, bear up to 100 flowers and present them over a long period of time—up to two months.

forma *variegata*

forma *bifolia*

Piperia

Piperia is a small genus of ten species found only in North America. The species are all superficially similar with few basal leaves, slender bracts on the stem, and a spike of very small, spurred flowers. With the exception of *Piperia unalascensis*, the genus is found strictly in western North America. There is much taxonomic history and confusion concerning *Piperia*, and it is yet to be all resolved. The genus *Piperia* was created in 1901 by Rydberg to accommodate those members of *Habenaria* and *Platanthera* with tuberoids, slightly swollen stems, slender spikes, flowers with the lips similar to the petals, and distribution primarily western North America. At that time there were only a few species in the new genus; subsequently several new species were described and the total is now at ten. In 1997 Bateman et al. returned the entire genus to *Platanthera* and made several new combinations on the basis of DNA analyses. Their view is not universally held as the morphology of those species in *Piperia* is distinctive from *Platanthera* in too many ways.

Piperia unalascensis (Sprengel) Rydberg
Alaskan piperia

forma *olympica* P.M. Brown—dwarf montane form
North American Native Orchid Journal 10: 37. 2004, type: Washington

Range: Alaska south to California, west to Montana, northern New Mexico, east to Ontario, Michigan, Quebec, Newfoundland
Within the prairies and Great Plains region:
Colorado, Montana, New Mexico, South Dakota, Wyoming: rare and local, more frequent westward
Plant: terrestrial; 10–50+ cm tall
Leaves: 2–4, weak and prostrate; usually withering at or before anthesis; 1.3–3.5 cm wide × 4.0–14.0 cm long
Flowers: 20–80+; arranged in a slender, tapering spike; translucent yellow-green, with a nocturnal, musky fragrance often lingering into the morning; sepals ovate-oblong, the lateral ones recurved; petals lanceolate-linear, projecting forward and approximate to the dorsal sepal; lip slender, similar to the petals, often descending; individual flower size ca. 2.5–3.0 mm, not including the 3–4 mm spur
Habitat: rocky woodlands and fens, usually on limestone
Flowering period: late June to early August

The **Alaskan piperia**, *Piperia unalascensis*, is the most widespread and frequently encountered member of the genus, and the only species of *Piperia* to be found in eastern North America. It is locally abundant in northeastern Michigan, Manitoulin Island, and the northern tip of the Bruce Peninsula, Ontario. Disjunct locations are found on the eastern Gaspé and Anticosti Island in Quebec and, most recently, in western Newfoundland. The Black Hills sites help to join the thread of the eastern and western North American populations. In all instances the plants are found growing in highly calcareous areas. The forma *olympica*, a dwarf habit from the Olympic Mountains of Washington, is occasionally found eastward in exposed subalpine and alpine areas.

In the recent trend to nest *Piperia* back into *Platanthera* the specific epithet *foetida* has been used. This is not correct as the epithet *unalascensis* predates *foetida* and the original genus used for *foetida*, *Herminium*, takes precedence over *Platanthera* in combination with *foetida* (IPNI).

Platanthera

The genus *Platanthera* comprises about 40 North American and Eurasian species, primarily of temperate climes, and is one of the major segregate genera traditionally placed by many botanists within *Habenaria*. It is the largest genus of orchids in the United States and Canada and, with fifteen species and two varieties found in the prairies and Great Plains region, the largest genus within our range. The plants are distinguished from *Habenaria* by their lack of basal rosettes as well as tubers or tuberoid roots. Many of the species have large, colorful, showy flowers in tall spikes or racemes. There are several sections to the genus but the most prominent is the section Blephariglottis, the fringed orchises. There are two groups within this section: those species with an entire, or unlobed, lip, and those with a three-parted lip.

Note: nearly all species of *Platanthera* can be found in both full sun and deeply shaded habitats. Plants in the sun tend to be shorter and have more densely flowered inflorescences and the leaves more upright, whereas those growing in shaded areas tend to be taller and have elongated, loosely flowered inflorescences with spreading leaves. The individual flower size remains the same, but the overall appearance of the plants can be markedly different, to the point that some observers initially think they have two different species!

Key to the fringed, bog, and rein orchises, *Platanthera*

1a lip with a distinct tubercle . . . 2
1b lip lacking a tubercle . . . 3

2a lip ovate; bracts generally equal to or shorter than the flowers . . . . . **southern tubercled orchis**, *Platanthera flava* var. *flava*, p. 144
2b lip oblong; bracts generally longer than the flowers . . . . . **northern tubercled orchis**, *Platanthera flava* var. *herbiola*, p. 146

3a leaves basal . . . 4
3b leaves cauline . . . 5

4a dorsal sepal and petals arching forward, lip triangular . . . . . **Hooker's orchis**, *Platanthera hookeri*, p. 148
4b petals spreading; lip linear . . . . . **pad-leaved orchis**, *Platanthera orbiculata*, p. 156

5a margin of lip fringed, lacerated, or erose . . . 6
5b margin of lip entire (none of the above) . . . 13

6a flowers orange or yellow . . . 7
6b flowers purple, green, or white . . . 8

7a spur longer than the lip . . . . . **orange fringed orchis,** *Platanthera ciliaris,* p. 136
7b spur distinctly shorter than the lip . . . . . **orange crested orchis,** *Platanthera cristata,* p. 138

8a lip entire; extirpated from the region . . . . . **northern white fringed orchis,** *Platanthera blephariglottis,* p. 134
8b lip divided (3-parted) . . . 9

9a flowers green . . . . . **green fringed orchis, ragged orchis,** *Platanthera lacera,* p. 152
9b flowers white or purple . . . 10

10a flowers creamy white . . . 11
10b flowers purple . . . 12

11a rostellum lobes parallel; spur 28–45 mm long; primarily east of the Mississippi River . . . . . **eastern prairie fringed orchis,** *Platanthera leucophaea,* p. 154
11b rostellum lobes spreading, spur 36–55 mm long . . . . . **western prairie fringed orchis,** *Platanthera praeclara,* p. 160

12a lip shallowly lacerate to ca. one-third the length of the lip . . . . . **small purple fringed orchis,** *Platanthera psycodes,* p. 163
12b lip merely erose . . . . . **purple fringeless orchis,** *Platanthera peramoena,* p. 158

13a flowers green . . . 14
13b flowers white . . . 16

14a spur tapered to clavate; equal to or shorter than the lip . . . 15
14b spur scrotiform or saccate; longer than the lip; very rare in Black Hills . . . . . **slender bog orchis,** *Platanthera stricta,* p. 166

15a lip yellowish to yellowish-green, rhombic-lanceolate . . . . . **northern green bog orchis,** *Platanthera aquilonis,* p. 131

15b lip whitish-green, lanceolate, usually obscurely rounded or slightly dilated at base . . . . . **green bog orchis,** *Platanthera huronensis,* p. 150

16a spur equal in length to the lip . . . . . **tall white northern bog orchis,** *Platanthera dilatata* var. *dilatata,* p. 140

16b spur shorter than the length of the lip; very rare in Black Hills . . . . . **bog candles,** *Platanthera dilatata* var. *albiflora,* p. 142

Platanthera aquilonis Sheviak

northern green bog orchis

forma *alba* (Light) P.M. Brown—albino form
  Lindleyana 4: 158. 1989, as *Platanthera hyperborea* forma *alba* M.H.S. Light, type: Quebec

Range: Alaska east to Newfoundland, south to California, New Mexico, and Iowa, east to Rhode Island
Within the prairies and Great Plains region:
Manitoba, Saskatchewan; Illinois, Iowa, Minnesota, Montana, Nebraska, North Dakota, South Dakota, Wisconsin: local to occasional throughout
Plant: terrestrial, 5–60 cm tall
Leaves: 2–4; cauline, linear-lanceolate, gradually reduced to bracts; 1–4 cm wide × 3–20 cm long
Flowers: 20–45; arranged in a loose to dense terminal spike; dorsal sepal obovate, arching; lateral sepals linear-oblong, spreading to recurved, petals rhombic-linear, somewhat enclosed within the dorsal sepal forming a hood; lip rhombic-lanceolate to lanceolate, descending, projecting or the apex caught within the tip of the dorsal sepal and petals; flowers yellow-green to whitish-green in cooler climes; lip usually a dull yellow-green; individual flower size 0.8 cm × 1.3 cm, not including the 0.2–0.5 cm clavate to somewhat cylindric spur; flowers are autogamous, with the pollinia rotating forward and falling out of the anther sacs, spilling onto the stigma of the column.
Habitat: open wet meadows, roadside ditches and seeps, fens, bogs and river gravels
Flowering period: late June through August

Plants formerly identified as *Platanthera hyperborea* in much of North America were described as a new species, *P. aquilonis*, by Sheviak in 1999 (common names are somewhat artificial within the green-flowered *Platanthera* species and it is best to just learn to use the Latin name). True *P. hyperborea* is known in North America only from Greenland, and all plants from other areas previously correctly assigned to that species are now *P. aquilonis*. In eastern and central North America the problem of the basic green *Platanthera* is not nearly as compounded as in western North America. Here in the prairies and plains region we have only two green-flowered species, *P. aquilonis* and *P. huronensis*, and the white-flowered *P. dilatata* (*P. stricta* is also known historically from the Black Hills of South Dakota). In western North America several other species occur as well. For many years both green-flowered species were simply referred to as *P. hyperborea* and usually in two varieties—var. *hyperborea* and var. *huronensis*. Unfortunately too many people, including authors, placed the smaller, slender, "poorly flowered" plants into *P. hyperborea* and the robust, lush-flowered plants into *P. huronensis*. That this was incorrect only compounded the problem. Sheviak's (1999) description of *P. aquilonis* greatly helped in solving this problem and recent work by Wallace (2003, 2004) validates both the origins and identifications of the three species. In a more simplistic form *P. aquilonis* may be differentiated from *P. huronensis* by the color and shape of the lip, position of the pollinia, and overall aspect of the plant. Range and habitat are also helpful but should not be relied upon too heavily. The following chart will help in comparing characteristics of the two species.

	Platanthera aquilonis	*Platanthera huronensis*
Lip/flower colour	Green flower with yellow lip	Whitish-green flower, lip not yellow
Lip shape	Sides nearly straight, angles rounded but not dilated	Lance shaped, base moderately to roundly dilated
Spur	Club shaped, approximately 3/4 length of lip, strongly forward curved	Slender to thickened cylindrical shape, equal in length to the lip, pendant to slightly forward curved
Scent	Absent	Fragrant
Pollination	Autogamous - pollinia rotate forward out of anther sacs, pollinia may break apart spilling the pollen	Autogamous - pollinia rotate forward out of anther sacs, pollinia may fragment spilling the pollen; *or may be* allogamous - pollinia remain in the anther sacs until removed by a pollinator
Anther sac position	Anther sacs low, widely diverging at the base and in close proximity at the apex	Anther sacs elevated, almost parallel, slightly diverging at the base and separated at the apex

Source: From Heshka (2003), reprinted by permission.

P. aquilonis   *P. huronensis*

Wild Orchids of the Prairies and Great Plains Region of North America

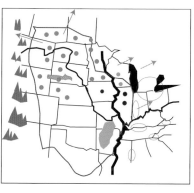

Platanthera blephariglottis (Willdenow) Lindley

northern white fringed orchis

forma *holopetala* (Lindley) P.M. Brown—entire-lip form
Genera and Species of Orchidaceous Plants 291. 1835, as *Platanthera holopetala* Lindley, type: Canada.

Range: Michigan east to Newfoundland, south to Georgia
Within the prairies and Great Plains region:
Illinois: a single historical record
Plant: terrestrial, 25–60 cm tall; plants much shorter in northern, exposed situations
Leaves: 2–4; cauline, lanceolate 2–4 cm wide × 8–20 cm long
Flowers: 20–45; arranged in a dense terminal raceme; sepals ovate, petals linear, enclosed within the sepals forming a hood; lip spatulate with a coarse, fringed margin or, in the forma *holopetala*, the margin slightly if at all fringed; lip narrowed to an obscure isthmus at the base; perianth pure white; individual flower size 1.75–2.0 cm, not including the 1.5–2.0 cm spur
Habitat: open wet meadows, roadside ditches and seeps, fens, and bogs
Flowering period: July into early August

The **northern white fringed orchis**, *Platanthera blephariglottis*, a large, showy member of the fringed-lipped section Blephlariglottis, is widespread throughout eastern North America and reaches the western and northern extreme of its distribution in Ontario. The eastern Illinois record represents a disjunct population that is considered extirpated. *Platanthera blephariglottis*, like several other species of the genus, is affected by successional habitats and often flowers best a few years after the disturbance of adjacent woody plant material by burning, mowing, or the like.

Platanthera ciliaris (Linnaeus) Lindley

orange fringed orchis

Range: southern Michigan east to Massachusetts, south to Florida and Texas
Within the prairies and Great Plains region:
Arkansas, Illinois, Louisiana, Missouri, Oklahoma, Texas: locally common throughout
Plant: terrestrial, 25–100+ cm tall
Leaves: 2–5; cauline, lanceolate 1–5 cm wide × 5–30 cm long, gradually reduced to bracts within the inflorescence
Flowers: 30–75; arranged in a dense terminal raceme; sepals ovate, petals linear, fringed at the tip, enclosed within the sepals forming a hood; lip ovate with a delicately fringed, filiform margin; perianth deep yellow to orange; the column tapering to a point; individual flower size 4 cm, not including the 2.5–3.5 cm spur, which is typically loosely descending
Habitat: open wet meadows, roadside ditches and seeps, pine flatwoods
Flowering period: late July and often to September in the south

The brilliant, deep yellow to orange plumes of the **orange fringed orchis**, *Platanthera ciliaris*, can be a meter in height. Scattered populations are well known in the southern and eastern portions of the prairies and plains and have been commonly found throughout the Mississippi Valley north to southwestern Michigan. *Platanthera ciliaris* has a preference for areas that stay moist to damp in the hot, dry days of summer; populations vary from year to year in how thrifty they are. Because neither of the white-flowered species of fringed orchises (*P. blephariglottis* and *P. conspicua*) currently occurs within the prairies and Great Plains, the hybrids *P.* ×*bicolor* (known from southwestern Michigan) and *P.* ×*lueri* will not be found. Hybrids with *P. cristata*, the **orange crested orchis**, are known as *P.* ×*channellii* and have been found in Arkansas, Texas, and Louisiana.

Platanthera cristata (Michaux) Lindley
orange crested orchis

> forma *straminea* P.M. Brown—pale yellow-flowered form
> *North American Native Orchid Journal* 1(1): 12. 1995, type: New Jersey

Range: southeastern Massachusetts south to Florida, and west to eastern Texas; primarily on the Coastal Plain
Within the prairies and Great Plains region:
Arkansas, Texas, Louisiana: rare and scattered throughout our area, primarily in areas of Coastal Plain influence
Plant: terrestrial, to 80 cm tall
Leaves: 2–4; cauline, lanceolate, 1–3 cm wide × 5–20 cm long, rapidly reduced to bracts within the inflorescence
Flowers: 30–80; arranged in a loose to dense terminal raceme; sepals ovate, petals spatulate with the margin of the apex crested, enclosed within the sepals forming a hood; lip long, triangular-ovate with a coarsely lacerate margin; perianth deep yellow to orange or, in the forma *straminea*, pale yellow; the column very short and the face flattened; individual flower size 5–7 mm, not including the 7 mm spur which is typically curved
Habitat: open wet meadows, roadside ditches and seeps, and pine flatwoods
Flowering period: late June through late September

A smaller, and perhaps more refined version of the **orange fringed orchis**, *Platanthera ciliaris*, the **orange crested orchis**, *P. cristata*, is a common species of the southeastern Coastal Plain and often occurs without other related species nearby. In the prairies and plains plants have occasionally been found in open pine flatwoods, although it can also be found in damp meadows, ditches, and roadside seeps. The raceme is usually about 2.5 cm in diameter and the spur is always shorter than the lip and typically curved forward, whereas *P. ciliaris* always has the loosely descending spur much longer than the lip, and the raceme is 4.5+ cm in diameter.

Hybrids with *Platanthera ciliaris*, the **orange fringed orchis**, are known as *P.* ×*channellii* and have been found in Arkansas, Texas, and Louisiana.

forma *straminea*

Platanthera dilatata (Pursh) Lindley var. *dilatata*

tall white northern bog orchis

Range: Alaska east to Newfoundland, south to California and New Mexico; Minnesota south to Indiana, Pennsylvania, and New England

Within the prairies and Great Plains region:
Manitoba; Illinois, Minnesota, Montana, Nebraska, South Dakota, Wisconsin: widespread and locally common northward

Plant: terrestrial, 25–100+ cm tall

Leaves: 4–12; cauline, passing into bracts on the stem, lanceolate, 4–7 cm wide × 15–30 cm long

Flowers: 20–100+; arranged in a dense terminal spike; dorsal sepal ovate, lateral sepals linear; petals linear-falcate, enclosed within the sepals forming a hood; lip linear-lanceolate, dilated at the base; perianth pure white; individual flower size 1.75–2.0 cm, not including the cylindric spur which is about equal to the length of the lip

Habitat: open wet meadows, roadside ditches and seeps, fens, and bogs

Flowering period: late June through much of August

The **tall white northern bog orchis**, *Platanthera dilatata*, is the showiest of the slender bog or rein orchids. Its tall, white, fragrant spikes may occur in great numbers throughout the range of the species. Size is variable and smaller plants often have only a few flowers. In the northern portion of its range mixed colonies of *P. dilatata*, *P. huronensis*, and/or *P. aquilonis* are not uncommon. Hybridization among these species is minimal today because of the evolution of self-pollinating breeding systems in the green-flowered species. Plants known as *P. ×media* were once thought to represent frequent hybrids between *P. dilatata* and *P. hyperborea;* although that idea is technically correct, they actually represent plants of *P. huronensis* (Sheviak, 1999; Wallace, 2004). *Platanthera dilatata* var. *albiflora* and var. *leucostachys* both occur primarily in western North America.

Wild Orchids of the Prairies and Great Plains Region of North America   141

Platanthera dilatata (Pursh) Lindley var. *albiflora* (Chamisso) Ledebour

bog candles

Range: Alaska south to Washington, east to Utah and Colorado
Within the prairies and Great Plains region:
South Dakota: very rare in the Black Hills
Plant: terrestrial, 25–100+ cm tall
Leaves: 4–12; cauline, passing into bracts on the stem, lanceolate, 4–7 cm wide × 15–30 cm long
Flowers: 20–100+; arranged in a dense terminal spike; dorsal sepal ovate, lateral sepals linear; petals linear-falcate, enclosed within the sepals forming a hood; lip 6–10 mm, linear-lanceolate, dilated at the base; perianth pure white; individual flower size 1.75–2.0 cm, not including the 2–7 mm cylindric spur which is markedly shorter than the lip
Habitat: open wet meadows, roadside ditches and seeps, fens, bogs
Flowering period: beginning in late June and continuing into August

The inclusion of this primarily western variety is based on collections made by Magrath from the Black Hills in 1972. His determination was based on the length of the spur. The nearest additional documented occurrence is several hundred miles to the west in central Wyoming.

The difference in the three varieties of *Platanthera dilatata* is usually based on spur length (Sheviak in FNA, 2002; Wallace, 2003). The difference in range as well may be related to specific pollinators. All three varieties have a spicy fragrance.

Platanthera flava (Linnaeus) Lindley var. *flava*
southern tubercled orchis

Range: southwestern Nova Scotia; Missouri east to Maryland, south to Florida and Texas
Within the prairies and Great Plains region:
Arkansas, Illinois, Louisiana, Missouri, Oklahoma, Texas: local and scattered throughout the area; extending well north and occasionally overlapping with the var. *herbiola*
Plant: terrestrial, 10–60 cm tall
Leaves: 2–4; cauline, nearly basal, lanceolate, 1–4 cm wide × 5–20 cm long, rapidly reduced to bracts within the inflorescence
Flowers: 10–40; arranged in a loose to dense terminal raceme; sepals and petals ovate, enclosed within the dorsal sepal forming a hood; lip ovate with a prominent tubercle in the center; perianth yellow-green; individual flower size 6–7 mm, not including the 8 mm spur
Habitat: open wet meadows, roadside ditches and seeps, swamps and shaded floodplains
Flowering period: late April in the south to July northward

Platanthera flava var. *flava*, the **southern tubercled orchis**, is equally at home in shaded, wet woodlands and streamsides, and in bright, sunny, open, damp roadsides. Although the flowers are identical in both habitats, the habit of the plant varies greatly. Those in shaded habitats tend to be tall and slender, with flowers spaced out along the stem, whereas those in sunnier habitats have flower spikes and leaves that are very compact and crowded. The flowers in the shade tend to be very green in color and those in the sun much more yellow, tending toward chartreuse. But, in both instances, that distinctive tubercled lip is always prominent.

Platanthera flava (Linnaeus) Lindley var. *herbiola* (R. Brown) Luer

northern tubercled orchis

forma *lutea* (Boivin) Whiting & Catling—yellow-flowered form
Naturaliste Canadien 109(2): 278. 1982, type: Ontario

Range: Minnesota east to Nova Scotia, south to Missouri and Georgia; south in the Appalachian Mountains
Within the prairies and Great Plains region:
Arkansas, Illinois, Iowa, Minnesota, Missouri, [Texas], Wisconsin: rare to local
Plant: terrestrial, 10–50 cm tall
Leaves: 3–5; cauline, nearly basal, lanceolate 2–5 cm wide × 8–20 cm long, rapidly reduced to bracts within the inflorescence
Flowers: 15–45; arranged in a loose to dense terminal raceme; sepals and petals ovate, enclosed within the dorsal sepal forming a hood; lip oblong with a prominent tubercle near the base and triangular lobes on either side; perianth grassy-green or, in the forma *lutea*, distinctly yellow; individual flower size 6–7 mm, not including the 5–7 mm spur
Habitat: open wet meadows, roadside ditches and seeps, swamps, and shaded floodplains
Flowering period: late May in the south through June northward

Platanthera flava var. *herbiola*, the **northern tubercled orchis,** is an obscure, grass-green, fragrant species once considered one of the rarer orchids in eastern North America. As the result of intensive field searches by many individuals we now know of some large, stable populations in several states. The northern variety of the **tubercled orchis** is at the southwestern extreme of its range in a reported site from northern Texas. Its sweet, perfume-like fragrance is often detected before the plant is actually seen. The nominate variety *flava* occurs primarily in the southeastern and south central United States with disjunct populations in southwestern Nova Scotia and southern Ontario.

Platanthera hookeri (Torrey) Lindley

Hooker's orchis

forma *abbreviata* (Fernald) P.M. Brown—dwarfed form
: *Rhodora* 35: 239, 1933 as *Habenaria hookeri* Torrey var. *abbreviata* Fernald, type: Newfoundland

forma *oblongifolia* (J.A. Paine) P.M. Brown—narrow-leaved form
: *Annual Report of the New York State Cabinet* 18: 135. 1865 as *Platanthera hookeri* var. *oblongifolia* J.A. Paine, type: New York.

Range: Minnesota east to Newfoundland, south to Iowa and New Jersey
Within the prairies and Great Plains region: **Manitoba; Illinois, Iowa, Minnesota, Wisconsin:** rare and local in limited areas
Plant: terrestrial, 10–50 cm tall or, in the forma *abbreviata*, to 18 cm tall, the inflorescence occupying nearly half the height; bracts absent from stem
Leaves: 2; basal, oblong-ovate up to 12 cm wide × 15 cm long or, in the forma *oblongifolia*, 5 cm wide × 12 cm long; light green above and pale beneath
Flowers: 5–25; arranged in a loose terminal raceme, individual flowers with the appearance of ice tongs; dorsal sepal concave, ovate and tapering to a point; lateral sepals lanceolate and strongly reflexed; petals linear-lanceolate, tapering, falcate, projecting forward; lip long-triangular curving upward at the tip; perianth lime-green; flower size ca. 2 cm × 3 cm not including the slender 1.5–2.5 cm spur; plants of the forma *abbreviata* smaller and more crowded in all aspects with the color nearly yellow-bronze
Habitat: rich deciduous and mixed woodlands
Flowering period: July

No native North American orchid is so curious in its appearance as the flowers of **Hooker's orchis**, *Platanthera hookeri*, and strikes some as looking like gargoyles or ice tongs. The lower lip curls upward, the dorsal sepal projects forward, and the petals spread wing-like, giving this appearance. Plants found in open woodlands with little competing ground cover often occur in colonies; although the plants are monochromatic—a decided shade of chartreuse—they are not difficult to spot. **Hooker's orchis** is one of three species of *Platanthera* that produces oval to round basal leaves. The other two, the **pad-leaved orchis**, *P. orbiculata*, and **Goldie's pad-leaved orchis**, *P. macrophylla* (the latter not present in the prairies and plains region), are both larger in overall dimensions, have much rounder leaves, and flower later than **Hooker's orchis**. Dwarfed plants found in the open heaths of western Newfoundland were originally designated as var. *abbreviata* by Fernald but are better treated as a form. They are very distinctive; the leaves are smaller and the flowers more crowded on the raceme and decidedly yellow with a bronze tinge. Given the habitat and western extreme of the range such plants are unlikely in the prairies and Great Plains region.

forma *oblongifolia*

Platanthera huronensis (Nuttall) Lindley

green bog orchis

Range: Alaska east to Newfoundland, south to California and Pennsylvania
Within the prairies and Great Plains region:
Manitoba; Minnesota, South Dakota, Wisconsin: widespread to local
Plant: terrestrial, 10–100+ cm tall
Leaves: 2–4; cauline, linear-lanceolate, gradually reduced to bracts; 1–6 cm wide × 5–30 cm long
Flowers: (8)20–75+; arranged in a loose to dense terminal spike; dorsal sepal obovate, arching; lateral sepals linear-oblong, spreading to recurved, petals ovate to lance-falcate, somewhat enclosed within the dorsal sepal forming a hood; lip lanceolate, descending, or the apex caught within the tip of the dorsal sepal and petals; sepals whitish-green, petals and lip pale greenish-white but markedly whiter than the sepals; individual flower size 0.8 cm × 1.3 cm, not including the 0.4–1.2 cm somewhat cylindric spur; flowers are autogamous, with the downward-pointing pollinia remaining in the anther sacs.
Habitat: open wet meadows, roadside ditches and seeps, fens, bogs, and river gravels
Flowering period: late June lasting well into August

Except for the prairies and plains region, the **green bog orchis**, *Platanthera huronensis*, is the most widespread and frequently encountered of all of the green-flowered bog or rein orchises found in northern North America. In our region it is confined to isolated disjunct sites and frequent only in the extreme northern portion of the range. In southern Manitoba adjacent to a massive site for the **western prairie fringed orchis**, *P. praeclara*, roadside ditches could easily be found with both *P. huronensis* and *P. aquilonis*. The tall spikes of *P. huronensis* are frequently found in a wide variety of habitats and, like many species of *Platanthera*, their habit varies with the habitat. Plants of open wet areas have densely flowered tall spikes with many flowers whereas those of the woodlands often have few-flowered, slender spikes. The flowers are usually intensely fragrant. Sheviak (in FNA, 2002) states that although hybrids with *P. dilatata* may occur, the name traditionally used for them, *P.* ×*media*, is actually a synonym for *P. huronensis*. See the chart on page 132 for a comparison with *P. aquilonis*.

Platanthera lacera (Michaux) G. Don
green fringed orchis, ragged orchis

Range: Manitoba east to Newfoundland, south to Texas and Georgia
Within the prairies and Great Plains region:
Manitoba; Arkansas, Illinois, Kansas, Minnesota, Missouri, Oklahoma, Texas, Wisconsin: scattered and local in most areas at the western limit of its range
Plant: terrestrial, 20–80 cm tall
Leaves: 3–6; cauline, lanceolate, keeled, 2.5–5.0 cm wide × 8.0–20.5(24.0) cm long, passing into bracts
Flowers: 12–40+, highly variable; arranged in a loose to dense terminal raceme; sepals obovate, the petals oblong, upright, usually with entire margins; lip three-parted and deeply lacerate; perianth varies from green to nearly yellow or white; individual flower size ca. 1.5–3.0 cm, not including the 1.6–2.3 cm spur, the orifice nearly square
Habitat: open wet meadows, roadside ditches and seeps, mountain meadows
Flowering period: late May through mid-August

The least conspicuous of the fringed orchises, the **green fringed orchis**, *Platanthera lacera*, is well scattered throughout the region. It can be found flowering during most of the summer in damp meadows, open wet woods, and roadside ditches. Flower color is highly variable in many shades of green, and some plants are nearly white. Further east the **green fringed orchis** is often found growing with the **small purple fringed orchis**, *P. psycodes*. Those plants whose flowers show a wash of lavender may represent hybrids with *P. psycodes* (*P. ×andrewsii*). *Platanthera ×hollandii*, which was described from a fen in southern Ontario is the hybrid with the **eastern prairie fringed orchis**, *P. leucophaea*. At least in theory, it might be found at the few sites where *P. leucophaea* and *P. lacera* are both found. Plants formerly known as *P. lacera* var. *terrae-novae* have proven to be *P. ×andrewsii* (Catling, 1997; Catling and Catling, 1994). See page 169 for more details.

Platanthera leucophaea (Nuttall) Lindley

eastern prairie fringed orchis

Range: Nebraska east to Ontario and Maine, south to Oklahoma, Louisiana, and Virginia
FEDERALLY LISTED AS THREATENED
Within the prairies and Great Plains region:
Illinois, Iowa, Louisiana, Missouri, Nebraska, Oklahoma, Wisconsin: very rare and local; presumed extirpated from Louisiana, Nebraska, and Oklahoma
Plant: terrestrial, 50–120 cm tall
Leaves: 2–5; cauline, lanceolate, keeled, 2.5–5.0 cm wide × 8.0–20.0 cm long, rapidly reduced to bracts within the inflorescence
Flowers: 12–27; arranged in a loose to dense terminal raceme; sepals ovate, the petals spatulate and broadly rounded, the margin finely serrate, partially enclosed with the sepals to form a cup-shaped hood; lip three-parted and deeply fringed; perianth cream shaded to whitish green; the sepals nearly entirely green; rostellum lobes parallel; individual flower size ca. 2–3 cm, not including the 3–5 cm spur, the orifice broadly crescent-shaped
Habitat: wet prairies, fens
Flowering period: June through much of July

Rapidly disappearing from unprotected sites, the **eastern prairie fringed orchis**, *Platanthera leucophaea*, is one of the most vulnerable species of orchids in North America. Extensive studies have been done over the years to attempt to understand its life cycle, response to burning, and rate of reproduction (Zettler et al., 2005). Many sites have only a few plants and even fewer that flower each year. Large stands are uncommon but do still exist on protected lands in southern Wisconsin and Illinois.

The inclusion of the **eastern prairie fringed orchis** in Louisiana is based on an old specimen labeled simply "Louisiana," with no further details. Sheviak (1987) gives more corroborating details that support this record. Living plants have not been seen here in recent history and the nearest historical population is a disjunct site several hundred miles away in Oklahoma. Little habitat for this species still exists in remnant prairies in northwestern Louisiana. The plants are unmistakable and could be confused only with the western prairie fringed orchis, *Platanthera praeclara*, which is equally as rare and with a more western distribution. The large, showy flowers are exceedingly fragrant and pollinated by sphinx moths.

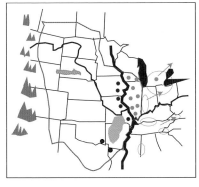

Platanthera orbiculata (Pursh) Lindley
pad-leaved orchis

forma *lehorsii* (Fernald) P.M. Brown—dwarfed form
> *Rhodora* 52: 61. 1950 as *Habenaria orbiculata* Pursh var. *lehorsii* Fernald, type: St-Pierre et Miquelon

forma *longifolia* (Clute) P.M. Brown—narrow-leaved form
> *American Botanist* (Binghamton) 7: 56. 1904 as *Habenaria orbiculata* Pursh var. *longifolia* Clute, type: Vermont

forma *pauciflora* (Jennings) P.M. Brown—few-flowered form
> *Journal of the Washington Academy of Science* 10: 453. 1920 as *Lysias orbiculata* Rydberg var. *pauciflora* Jennings, type: Ontario

forma *trifolia* (Mousley) P.M. Brown—3-leaved form
> *Orchid Review*. 42: 112. 1934 as *Habenaria orbiculata* Pursh var. *trifolia* Mousley, type: Quebec

Range: southeastern Alaska, British Columbia east to Newfoundland, south to Washington and Maryland, south in the Appalachian Mountains to North Carolina [Georgia]

Within the prairies and Great Plains region:
Manitoba; Illinois, Montana, South Dakota, Wyoming: rare to local, but often occurring in large colonies

Plants: 12–75 cm tall; stem bracts present

Leaves: 2, basal, lying on the ground, or, in the forma *trifolia*, 3, the third leaf on the stem; ovate up to (4)10–20 cm wide × (5)12–22 cm long or, in the forma *longifolia*, oblong to nearly linear; dark green above and pale beneath

Flowers: 8–14 or, in the forma *pauciflora*, 3–5 on shorter plants; arranged in a loose terminal raceme; dorsal sepal broadly ovate tapering to a point and concave; lateral sepals ovate-falcate and strongly reflexed; petals linear-lanceolate, tapering, falcate, erect; lip linear-oblong, descending or rarely recurved; sepals greenish-white; petals and lip whiter; flower size ca. 3 cm × 5 cm not including the slender 1.4–1.7 cm spur; plants of the forma *lehorsii* from St.-Pierre & Miquelon Islands and western Newfoundland, smaller and more crowded in all aspects with the color approaching whitish-bronze

Habitat: mixed woodlands, forma *lehorsii* in exposed sites on headlands

Flowering period: primarily July and into early August

Note: *Platanthera macrophylla*, **Goldie's pad-leaved orchis**, does not occur within the prairies and plains region treated in this book, but it is essential to understand the relationship of the two species.

The two pad-leaved orchises, *Platanthera orbiculata* and *P. macrophylla*, presented, until recently, one of the most misunderstood species pairs in North American orchids. The work of Allan and Joyce Reddoch (1993) clearly revalidated Goldie's pad-leaved orchis, *P. macrophylla*, as a full species and helped to delineate the bounds of the **pad-leaved orchis**, *P. orbiculata*. Overall size does not matter except in the length of the spur. Spurs of *P. orbiculata* are 1.4–2.7 cm long, the lesser measurement usually found in plants of forma *lehorsii* and forma *pauciflora*. Mixed colonies of the two species are not uncommon in the eastern portion of the range but as one progresses westward, *P. macrophylla* becomes less common and pure stands of *P. orbiculata* are the norm. The large, round leaves are always distinctive and the colonies in the Black Hills of South Dakota and Wyoming are well documented (Hornbeck et al., 2003).

Platanthera peramoena (A. Gray) A. Gray

purple fringeless orchis

> forma *doddsiae* P.M. Brown—white-flowered form
> North American Native Orchid Journal 8: 30–31. fig. 2002, type: Missouri

Range: Missouri east to New Jersey, south to Mississippi and Georgia

Within the prairies and Great Plains region:
Arkansas, Illinois, Missouri, Oklahoma: occasional and local, but often in large colonies

Plant: terrestrial, 75–105 cm tall

Leaves: 2–5; cauline, elliptic-lanceolate, keeled, 2.5–5.0 cm wide × 8.0–14.0 cm long, gradually reduced to bracts within the inflorescence

Flowers: 27–50+; arranged in a dense terminal raceme; sepals elliptic, reflexed; petals spatulate and rounded, the margin finely serrate, partially enclosed with the dorsal sepal to form a hood; lip three-parted, apically notched and faintly erose; perianth a brilliant rosy-purple or, in the forma *doddsiae*, pure white; individual flower size ca. 2–3 cm, not including the 2.5–3.0 cm spur, the orifice rounded

Habitat: damp meadows, low wet woods, streamsides, roadside ditches

Flowering period: midsummer; late June to early August

One of the tallest and most striking of the fringed orchises, *Platanthera peramoena*, the **purple fringeless orchis**, is a characteristic species of the east-central United States. Not necessarily common anywhere, it can be locally abundant. In the Midwest it is most frequently seen in southern Illinois, Missouri, and northeastern Arkansas. In deeply shaded habitats short (to 30 cm) plants with few flowers may be found. Encountering any of these plants in flower is a real delight and, despite careful searching, the white-flowered form, forma *doddsiae*, has been reported only once, from Missouri (Summers, 1998; Brown, 2003). Not easy to overlook, it should be carefully sought elsewhere.

forma *doddsiae*

Wild Orchids of the Prairies and Great Plains Region of North America   159

Platanthera praeclara Sheviak & Bowles
western prairie fringed orchis

Range: Manitoba south to Oklahoma
FEDERALLY LISTED AS THREATENED
Within the prairies and Great Plains region: Manitoba; Iowa, Kansas, Missouri, Nebraska, North Dakota, Oklahoma, South Dakota, Wyoming: very rare and local; presumed extirpated from Wyoming
Plant: terrestrial, 40–97 cm tall
Leaves: 2–5; cauline, lanceolate, keeled, 3.5–5.0 cm wide × 12–26 cm long, rapidly reduced to bracts within the inflorescence
Flowers: 10–24; arranged in a loose to dense terminal raceme; sepals ovate, the petals flabellate and flattened on top, the margin finely lacerate, partially enclosed within the sepals to form a cup-shaped hood; lip three-parted and deeply fringed; perianth creamy-white; rostellum lobes spreading; individual flower size ca. 3–5 cm, not including the 3.6–6.4 cm spur, the orifice rounded
Habitat: wet prairies, fens, open plains
Flowering period: June through much of July

Platanthera praeclara, the **western prairie fringed orchis,** is the orchid showpiece of the Great Plains. Although the plant ranges from Manitoba south to Oklahoma there are few protected localities and many of these are on federal lands. As often happens, the largest extant site is at the northern limit of the range in southern Manitoba, with several thousands of plants flowering annually—some years to 20,000! It is a magnificent sight/site to behold! With tall stems to a meter in height and inflorescences larger than a man's hand, it is one of the most breathtaking experiences to behold in the North American orchid world.

Concerning the western limit in Wyoming, Chuck Sheviak (pers. comm.) has stated,

> The Wyoming record is based on a single collection by John Fremont, made in 1842 on an early expedition. A Fremont specimen at the Oakes Ames Orchid Herbarium at the Harvard University Herbaria is labeled simply "Platte Bottom," but Bill Jennings was familiar with the history of the expedition and provided additional details from the published expedition report. The species was collected on 27 July 1842, in the "Black Hills." On this date the expedition was camped along the North Platte River in the vicinity of present day Casper, and the "Black Hills" of the report are the present-day Laramie Range, not those of South Dakota. This looks like a very good record, and I

think it is unambiguous. But there is no evidence that the plant occurs in the state now, and it is virtually certain that it doesn't.

In July 2003 Christie Borkowsky discovered lavender-pink flowered plants within a population of *Platanthera praeclara* in the Tall Grass Prairie Preserve in southern Manitoba. A photograph by Lorne Heshka, which originally appeared in *Wild Orchids of Manitoba* (Ames et al., 2005, p. 40), is included here. At the original location were three plants of this interesting color—all nearly identical to Heshka's photograph. Then, within a few days, a similar one was found at a different location—approximately 1.5 km to the south. In 2005 a single plant was found by Chris Friesen, a University of Manitoba Master's student, while carrying out research on the western prairie fringed orchis pollinators. The 2005 plant was approximately 100 m from the location of the original plants seen in 2003. Speculation regarding the origin of these plants includes the obvious, although undocumented, possibility of a hybrid between *P. praeclara* and *P. psycodes*. The nearest known population of *P. psycodes* in Manitoba is ca. 120 km east of the Tall Grass Prairie Preserve. However, the Minnesota distribution map in *Orchids of Minnesota* (Smith, 1993) indicates that an herbarium specimen of *P. psycodes* has been collected in northern Minnesota immediately south of the Tall Grass Prairie Preserve. The latter record

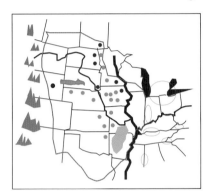

lends more support to the possibility of a hybrid (Heshka, 2005, pers. comm.)

The **western prairie fringed orchis** is federally listed as threatened and has had perhaps more efforts directed toward its status and recovery than any other orchid in North America (CPC Plant Profile 9293).

pink-flowered plant

Platanthera psycodes (Linnaeus) Lindley
small purple fringed orchis

forma *albiflora* (R. Hoffman) Whiting & Catling—white-flowered form
  Proceedings of the Boston Society of Natural History 36: 248. 1922, as *Habenaria psycodes* (Linnaeus) Sprengel forma *albiflora* R. Hoffman, type: Massachusetts

forma *ecalcarata* (Bryan) P.M. Brown—spurless form
  Annals of the Missouri Botanical Garden 4: 38. 1917, as *Habenaria psycodes* (Linnaeus) Sprengel var. *ecalcarata* Bryan, type: Michigan

forma *fernaldii* (Rousseau & Rouleau) P.M. Brown—dwarf from
  Bulletin du Jardin Botanique de l'État 27: 370. 1957, as *Habenaria psycodes* (Linnaeus) Sprengel forma *fernaldii* Rousseau & Rouleau, type: Quebec

forma *rosea* P.M. Brown—pink-flowered form
  North American Native Orchid Journal 1(4): 289. 1995, type: Vermont

forma *varians* (Bryan) P.M. Brown—entire-lip form
  Annals of the Missouri Botanical Garden 4: 37. 1917, as *Habenaria psycodes* (Linnaeus) Sprengel var. *varians* Bryan, type: Michigan

Range: Ontario east to Newfoundland, south to West Virginia and New Jersey; south in the Appalachian Mountains to Georgia
Within the prairies and Great Plains region:
Manitoba; Illinois, Iowa, Minnesota, [Nebraska], **Wisconsin:** rare and local at the western extreme of the range
Plant: terrestrial, to 90 cm tall or, in the forma *fernaldii*, only 10–15 cm tall
Leaves: 2–6; cauline, lanceolate, keeled 1.5–7.0 cm wide × 8.0–24.0 cm long, gradually reduced to bracts within the inflorescence
Flowers: 30–125 or, in the forma *fernaldii* only 8–20; arranged in a loose to dense terminal raceme usually 2.5–3.0 cm in diameter with flowers opening successively, i.e., the lower ones usually withering before the upper ones have opened, giving the inflorescence a conical appearance; sepals elliptic, petals obovate with finely dentate margins; lip 3-parted with a finely fringed margin usually to less than one-third the depth of the lip, or in the forma *varians* the margin essentially entire; perianth various shades of purple from pale lavender to deep, rich rosy-magenta or, in the forma *albiflora*, white or, in the forma *rosea*, a pale pink; individual flower size 0.5–1.5 cm, not including the 1.2–1.8 cm spur, or in the forma *ecalcarata,* the spur lacking; spur orifice likened to a transverse dumbbell
Habitat: open wet meadows, roadside ditches and seeps, mountain meadows
Flowering period: late June to early July

Note: Although the **large purple fringed orchis**, *Platanthera grandiflora*, does not occur within the prairies and plains region it is essential to understand the relationships between the two species.

The common names **large** and **small purple fringed orchis** are very misleading as the **small purple fringed orchis,** *Platanthera psycodes,* can often be "larger" than the **large purple fringed orchis,** *P. grandiflora*. The **small purple fringed orchis** is usually both taller and more floriferous than *P. grandiflora*, although the individual flowers are smaller. The **small purple fringed orchis** is also widespread throughout much of central and northeastern North America. This tall, slender species is at home in open meadows as well as along wooded streamsides. It usually occurs in small numbers, but is rarely found as a single plant. Because it is reaching the western extreme of its range in Manitoba, Illinois, and (formerly) Missouri, variation may be minimal. The Nebraska report is of a plant found in a forest plantation and is fragmentary (Kaul, 1986). Hybrids with *P. lacera* are known as *P.* ×*andrewsii*.

forma *albiflora*

forma *rosea*

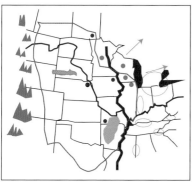

Platanthera stricta Lindley

slender bog orchis

Range: Alaska east to Alberta, south to California and Wyoming
Within the prairies and Great Plains region:
South Dakota (Black Hills): very rare
Plant: terrestrial, 30–100+ cm tall
Leaves: 4–10; cauline, oblong to obtuse, gradually reduced to bracts; 2.0–3.5 cm wide × 10.0–30.0 cm long
Flowers: 20–40+; arranged in a loosely to densely flowered spike; dorsal sepal obovate-triangular, arching; lateral sepals oblong, spreading, slightly twisted or barely recurved; petals falcate-acuminate and enclosed within the dorsal sepal forming a hood; lip lanceolate, obtuse, descending, 4–8 mm long; flowers green throughout; individual flower size 0.75–1.0 cm × 1.5–2.0 cm, not including the 0.2–0.4 cm scrotiform or strongly clavate spur that is much shorter than the length of the lip
Habitat: open wet meadows, roadside ditches and seeps, fens, bogs, and river gravels
Flowering period: late June to August

Although the **slender bog orchis,** *Platanthera stricta*, may be the most widely distributed green-flowered *Platanthera* species in northwest North America it reaches its eastern limit in the Black Hills of South Dakota. Magrath originally collected specimens of *Platanthera stricta*, as *Habenaria saccata*, in 1967. The specimens at KANU have been annotated and reannotated by several people as *Habenaria* or *Platanthera hyperborea* or *P. aquilonis*. Recent examination of these specimens and the duplicate at New York Botanical Garden clearly show that they are *P. stricta* as Magrath originally had determined. Plants have not been noted in the Black Hills since Magrath's original collections. In theory two possible hybrids could occur in this area. *Platanthera* ×*estesii* with *P. dilatata* var. *albiflora* and *P.* ×*correllii* with *P. aquilonis* (as *P. hyperborea*).

Wild Orchids of the Prairies and Great Plains Region of North America

Hybrids:
Platanthera ×*andrewsii* (Niles) Luer
Andrews' hybrid fringed orchis
(*P. lacera* × *P. psycodes*)
> *Bog-trotting for Orchids*. 258. 1904, as *Habenaria* ×*andrewsii*, type: Vermont

Not uncommon throughout most of the range of the two parents, but because of the rarity of *Platanthera psycodes* in the Midwest the hybrid is less likely.

Platanthera ×*channellii* Folsom
Channell's fringed orchis
(*P. ciliaris* × *P. cristata*)
> *Orquidea* (Mex.) 9(2): 344. 1984, type: Alabama

This hybrid and the species *Platanthera chapmanii* that is found further south can be difficult to tell apart. Plants found that are intermediate between *P. ciliaris* and *P. cristata* in the southern prairies and Great Plains region are almost certain to be *P.* ×*channellii*.

Two interesting hybrids with *Platanthera leucophaea* have recently been described from southwestern Ontario: *Platanthera ×hollandii*, with *P. lacera* (Catling et al., 1999), and *P. ×reznicekii*, with *P. psycodes* (Catling and Brownell, 1999). These sites are the only documented instances of the two hybrid combinations having been found. Because of the rarity of the parents in the prairies and Great Plains the likelihood of the hybrids is lessened. The best field marks are *P. leucophaea* with an overlay of pink or pink tinges in the lip, indicating possible hybridization with *P. psycodes*, or *P. leucophaea* with deeply cut or lacerate segments, indicating possible hybridization with *P. lacera*. The petals and sepals are also intermediate with the respective parents. Remember, though, that hybrids usually favor one parent or the other, so differences may be slight. See *Wild Orchids of the Canadian Maritimes and Northern Great Lakes Region* (Brown and Folsom, 2006) for more information on these two hybrids.

Two additional hybrids that involve the western *Platanthera stricta* and theoretically might be found in the Black Hills are:

Platanthera ×correllii

P. aquilonis (as *P. hyperborea*)× *P. stricta*.

Platanthera ×estesii

P. dilatata var. *albiflora* × *P. stricta*

See *Wild Orchids of the Pacific Northwest and Canadian Rockies* (Brown and Folsom, 2006) for more information on these two hybrids.

Pogonia

This is a small genus of only three species, found in both Asia and North America. Formerly the genus included those species, among others, that are now treated in *Triphora*, *Isotria*, and *Cleistes*, although some current authors are again including *Isotria* and *Cleistes*. We have only a single species in the United States and Canada, *Pogonia ophioglossoides*, which has one of the broadest ranges of any North American orchid, throughout all of eastern North America, north of Mexico.

Pogonia ophioglossoides (Linnaeus) Ker-Gawler
rose pogonia; snakemouth orchid

forma *albiflora* Rand & Redfield—white-flowered form
Flora of Mount Desert Island 152. 1894, type: Maine
forma *brachypogon* (Fernald) P.M. Brown—short-bearded form
Rhodora (http://www.us.ipni.org/ipni/PublicationServlet?id=1065-2&query_type=by_id) 23:245. 1922, as Pogonia ophioglossoides (Linnaeus) Ker-Gawler var. brachypogon Fernald, type: Nova Scotia

Range: Manitoba east to Newfoundland, south to Texas and Florida
Within the prairies and Great Plains region:
Manitoba; Arkansas, Illinois, Louisiana, Minnesota, Missouri, North Dakota, Oklahoma, Texas, Wisconsin: locally common especially in the eastern portion of the prairie states
Plant: terrestrial, 8–35 cm tall
Leaves: 1, rarely 2; cauline, ovate, placed midway on the stem, 6–10 cm × 2 cm
Flowers: 1–3 (4); terminal; sepals and petals similar, lanceolate; the sepals wide-spreading; lip to 2 cm long, spatulate with a deeply fringed margin and bright yellow beard or, in the forma *brachypogon*, the beard reduced to a few colorless knobs; perianth from light to dark, rosy-pink or lavender or, in the forma *albiflora*, pure white; individual flower size ca. 4 cm
Habitat: moist meadows, open bogs and prairies, roadside ditches, and sphagnous seeps
Flowering period: May and June throughout most of the range and into July in the north

From Newfoundland to Florida and westward to the Mississippi Valley and eastern limits of the prairies and Great Plains region, this little jewel adorns open bogs and meadows, roadside ditches, borrow pits, and sphagnous seeps. Its pink flowers and broad spatulate lip may remind one of roseate spoonbills. Although variable in color, form, and size, the rose pogonias are a true herald of a solid spring of wild orchids. They are often seen in the company of grass-pinks, *Calopogon*, and ladies'-tresses, *Spiranthes*. Color and form vary greatly from colony to colony. It is not unusual to find plants with the petals and sepals very narrow and, within the same colony, individuals with the sepals and petals broad and rounded. The coloring of the flowers can range from pale lilac to intense magenta and there may occasionally be white-flowered plants among them. Although most plants have solitary flowers it can be interesting, especially in a large population, to search for those with multiple flowers. Stems with two flowers are not that unusual and, upon rare occasion, a three-flowered stem will be seen. Only once has the author seen a stem with four flowers! The unusual forma *brachypogon*, although described originally from Nova Scotia, has recently been seen, or more correctly detected, throughout the range of the species.

forma *albiflora*

Wild Orchids of the Prairies and Great Plains Region of North America   173

Spiranthes

Spiranthes is a cosmopolitan genus of about fifty species. Treated in the strictest sense it is one of the most easily recognized genera but with some of the more difficult plants to identify to species. The relatively slender, often twisted, stems and spikes of small white or creamy-yellow (or pink in *S. sinensis*) flowers are universally recognizable. In the United States and Canada we have twenty-four species, with thirteen species and two additional varieties found in the prairies and Great Plains region of North America.

Note: *Spiranthes cernua* is a compilospecies, with gene flow from several other species, depending on the plant's geographic location. Occasionally these plants prove problematic to key out. An unusual, nearly yellow-flowered cleistogamous/peloric race occurs in East Texas and western Louisiana.

Key to the ladies'-tresses, *Spiranthes*

1a leaves present at flowering time . . . 2
1b leaves lacking at flowering time . . . 15

2a flowers uniformly white or cream-colored (the lip may be a deeper cream color than the petals and sepals) . . . 3
2b flowers with green or yellow on the lip . . . 10

3a late spring or summer flowering . . . . . **grass-leaved ladies'-tresses,** *Spiranthes vernalis*, p. 204
3b summer or autumn flowering . . . 4

4a summer flowering . . . 5
4b autumn flowering . . . 7

5a lip creamy white to cream-colored . . . 6
5b lip with a green throat . . . . . **northern slender ladies'-tresses,** *Spiranthes lacera* var. *lacera*, p. 182

6a lip with a slight constriction in the middle; eastern Wyoming, Colorado and western Nebraska . . . . . **Ute ladies'-tresses,** *Spiranthes diluvialis*, p. 180
6b lip not constricted and with a lacerate margin; southern Arkansas; Louisiana and eastern Texas . . . . . **lace-lipped ladies'-tresses,** *Spiranthes laciniata*, p. 186

7a Flowers small, less than 5 mm . . . 8
7b Flowers greater than 5 mm long . . . 10

8a plants of woodlands and other shady habitats . . . 9
8b plants of open grasslands; flowers, ca. 5 mm long, usually arranged in a single rank,; open oak savannas of Wisconsin . . . . . **Case's ladies'-tresses,** *Spiranthes casei*, p. 176

9a flowers fully sexual with a rostellum; rare plants of the southern states . . . . . **southern oval ladies'-tresses,** *Spiranthes ovalis* var. *ovalis*, p. 194
9b flowers lacking a rostellum; widespread plants of central states . . . . . **northern oval ladies'-tresses,** *Spiranthes ovalis* var. *erostellata*, p. 196

10a lip constricted in the middle, panduriform . . . . . **hooded ladies'-tresses,** *Spiranthes romanzoffiana*, p. 198
10b lip not constricted in the middle . . . 11

11a flowers white or cream . . . 12
11b flowers with green or yellow on the lip . . . 14

12a lateral sepals and petals appressed . . . . . **nodding ladies'-tresses,** *Spiranthes cernua*, p. 178
12b lateral sepals and petals divergent . . . 13

13a lateral sepals arched and incurved . . . . . **Great Plains ladies'-tresses,** *Spiranthes magnicamporum*, p. 190
13b lateral sepals projecting forward . . . . . **fragrant ladies'-tresses,** *Spiranthes odorata*, p. 192

14a flowers light green to pale creamy-green with green veins . . . . . **woodland ladies'-tresses,** *Spiranthes sylvatica*, p. 200
14b flowers white, the lip with a yellow central portion . . . . . **shining ladies'-tresses,** *Spiranthes lucida*, p. 188

15a flowers, including lip, white or cream-colored; autumn flowering . . . 16
15b flowers white, the central portion of the lip green . . . . . **southern slender ladies'-tresses,** *Spiranthes lacera* var. *gracilis*, p. 184

16a summer flowering; flowers as arranged in a single rank . . . . . **little ladies'-tresses,** *Spiranthes tuberosa*, p. 202
16b autumn flowering; flowers as arranged in multiple ranks . . . 17

17a lateral sepals arched and incurved . . . . . **Great Plains ladies'-tresses,** *Spiranthes magnicamporum*, p. 190
17b lateral sepals and petals appressed . . . . . **nodding ladies'-tresses,** *Spiranthes cernua*, p. 178

Spiranthes casei Catling & Cruise var. *casei*

Case's ladies'-tresses

Range: Ontario east to Nova Scotia, south to Wisconsin, northern Pennsylvania, and western Maine
Within the prairies and Great Plains region:
Wisconsin: local
Plants: terrestrial, 8–50 cm tall, sparsely pubescent
Leaves: 3–5; appearing basal or on the lower portion of the stem; linear-oblanceolate, up to 2 cm wide × 20 cm long; ascending to spreading; leaves present at anthesis
Flowers: 10–50; in a spike, loosely spiraled with 5 or more flowers per cycle, nodding from the base of the perianth; floral bracts green; sepals lanceolate; lateral sepals slightly spreading; petals ovate to oblanceolate; perianth ivory or greenish-white; lip oblong, 5.0–7.5 mm, the central portion often a deeper creamy yellow, with thin, fringed margins, the apex truncate; overall flower size 6–9 mm long
Habitat: dry open sites usually on the Canadian Shield in shaley soils, road scrapes, or thin-soil grasslands
Flowering period: late August through September

Although plants of **Case's ladies'-tresses**, *Spiranthes casei*, had been known for many years, it was only in 1974 that they were described as a species; prior to this these plants were often known as the "northern (*Spiranthes*) *vernalis*," a species that grows considerably further to the south. For a short time plants of *Spiranthes casei* were also known as *S. intermedia*, again a completely different plant that is actually a hybrid between the **grass-leaved ladies'-tresses**, *S. vernalis* and the **southern slender ladies'-tresses**, *S. lacera* var. *gracilis*. With the publication of *Spiranthes casei* by Catling and Cruise in 1974, the mystery of the "northern *vernalis*" was solved. The species was named in honor of Fred Case, orchidist and botanist from Michigan and author of *Orchids of the Western Great Lakes Region*.

Plants of *Spiranthes casei*, like most other species of *Spiranthes*, vary greatly in size, vigor, and number of flowers. The small, nodding, partially open flowers in a single rank make them reasonably easy to spot and, if growing among other species of *Spiranthes*, they are very distinctive. In the prairies and Great Plains region *S. casei* is confined to the oak savannas of western Wisconsin. This habitat borders on the prairie and is a distinct and somewhat specialized area that harbors several orchid species (Weber, 1996).

Wild Orchids of the Prairies and Great Plains Region of North America   177

Spiranthes cernua (Linnaeus) L.C. Richard
nodding ladies'-tresses

Range: South Dakota east to Nova Scotia, south to Texas and Florida
Within the prairies and Great Plains region:
Arkansas, Illinois, Iowa, Kansas, Louisiana, Minnesota, Missouri, Nebraska, Oklahoma, South Dakota, Texas, Wisconsin: widespread and often frequent in all but Louisiana
Plant: terrestrial, 10–50 cm tall
Leaves: 3–5; appearing basal or on the lower portion of the stem; linear-oblanceolate, up to 2 cm wide × 26 cm long; ascending to spreading; leaves usually present at anthesis in most races, although in the Deep South and some prairie races they are absent
Flowers: 10–50; in a spike, tightly to loosely spiraled with 5 or more flowers per cycle, nodding from the base of the perianth or rarely ascending; sepals and petals similar, lanceolate; perianth white, ivory, or in some races greenish; lip oblong, broad at the apex, the central portion of the lip, in some races, creamy-yellow or green; the sepals approximate and extending forward, sometimes arching above the flower; individual flower size 0.6–10.5 mm
Habitat: wet to dry open sites, lightly wooded areas, moist grassy roadsides, pine flatwoods, etc.
Flowering period: autumn flowering; September through November (December)

Of all of our native orchids in North America the **nodding ladies'-tresses**, *Spiranthes cernua*, is the most difficult to give a simple, concise description and narrative. Because it is a compilospecies—with gene flow from several different similar species—plants in different geographic areas have strong resemblances to the basic diploid species contributing that unidirectional gene flow. In other words, those plants growing in close proximity to the **fragrant ladies'-tresses**, *S. odorata*, in the southern states have a greater resemblance to *S. odorata*, whereas those from the drier northern and western portions of its range bear strong affinities to the **Great Plains ladies'-tresses**, *S. magnicamporum*. Plants found in the northeastern states would have gene flow from the **yellow ladies'-tresses**, *S. ochroleuca*. In many areas, especially away from the Coastal Plain, this is the common autumn flowering *Spiranthes*. Those who are seriously interested in learning and determining the various races in their areas are urged to read Sheviak's entries for *S. cernua*, *S. odorata*, *S. magnicamporum*, and *S. ochroleuca* in *Flora of North America* (FNA, 2002). Many plants of *S. cernua* seen throughout the prairies and Great Plains are fairly similar to each other. Only in areas where the monoembryonic diploid species occur will there be plants that strongly emphasize the unidirectional gene flow. Sheviak (1982) illustrates six different "ecotypes" of *S. cernua* from the prairies and Great Plains states.

Spiranthes cernua representing unidirectional gene flow from *S. magnicamporum*

cleistogamous form

Wild Orchids of the Prairies and Great Plains Region of North America

Spiranthes diluvialis Sheviak
Ute ladies'-tresses

Range: Washington east to Montana, south to Utah, Nevada, Colorado, Wyoming, and Nebraska
Within the prairies and Great Plains region:
Colorado, Nebraska, Wyoming: very rare and local; disjunct or at the northern limit of the range
Plant: terrestrial, 18–65 cm tall
Leaves: 3–6; basal and extending up the lower fourth of the stem; linear-lanceolate, up to 1.5 cm wide × 28.0 cm long, passing into a few slender bracts; present at flowering time
Flowers: 10–60; in a dense spiral with 3–4 flowers per cycle; ascending; sepals and petals similar, creamy white or ivory, lanceolate, acuminate; lateral sepals free from the petals and lip, the lateral sepals loosely incurved and often positioned above the rest of the flower; the lip often with a yellower central portion, oblong-lanceolate, somewhat panduriform, the broadened margin entire or finely lacerate; individual flower size 0.9–1.2 cm
Habitat: rocky riverbanks, seeps, meadows
Flowering period: mid-July into late August

Originally described by Sheviak in 1984, the **Ute ladies'-tresses**, *Spiranthes diluvialis*, became the second of a quartet of new North American species of *Spiranthes* published by him within a ten-year period. *Spiranthes diluvialis* is of special interest as it is an amphidiploid hybrid product of the **hooded ladies'-tresses**, *S. romanzoffiana*, and the **Great Plains ladies'-tresses**, *S. magnicamporum*. Clearly exhibiting characters of both ancestral parents, the evolved new species has colonized several areas within the Rocky Mountains and adjacent areas. It is found growing primarily in wetlands and often is the only *Spiranthes* in the area. Originally described from Colorado, the species is now known from the western edge of the Great Plains in Nebraska and adjacent Wyoming as well as Colorado.

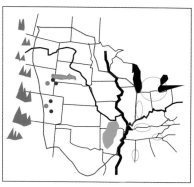

Wild Orchids of the Prairies and Great Plains Region of North America 181

Spiranthes lacera Rafinesque var. *lacera*
northern slender ladies'-tresses

Range: Alberta east to Nova Scotia, south to Missouri and Virginia
Within the prairies and Great Plains region:
Manitoba; Arkansas Illinois, Iowa, Kansas, Minnesota, Missouri, Wisconsin: locally frequent, more so in the northern part of the range becoming rare southward
Plant: terrestrial, 15–65 cm tall; pubescent
Leaves: 2–4; ovate, dark green, 1–2 cm wide × 2–5 cm long, usually present at flowering time
Flowers: 10–35; in a single rank, in a dense spiral; sepals and petals similar, elliptic; perianth white; lip oblong, with the apex rounded; central portion green with a clearly defined crisp apron; the lower flowers spaced out from those above; individual flower size 4.0–7.5 mm
Habitat: dry to moist meadows, grassy roadsides, cemeteries, open sandy areas in woodlands, lawns, old fields, oak savannahs
Flowering period: July to early October

The differences between the **northern slender ladies'-tresses**, *Spiranthes lacera* var. *lacera*, and the more southerly var. *gracilis* are not great, but the more northern of the two has the lower flowers well spaced out on the inflorescence and they appear to be much smaller because of the position of the sepals.

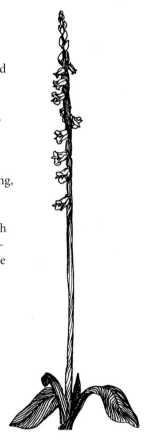

Wild Orchids of the Prairies and Great Plains Region of North America   183

Spiranthes lacera Rafinesque var. *gracilis* (Bigelow) Luer

southern slender ladies'-tresses

Range: Kansas, Michigan east to Maine, south to Texas and Georgia
Within the prairies and Great Plains region:
Arkansas, Illinois, Iowa, Kansas, Louisiana, Minnesota, Missouri, Oklahoma, Texas, Wisconsin: widespread and relatively common in both open and lightly wooded areas
Plant: terrestrial, 15–65 cm tall, glabrous to sparsely pubescent
Leaves: 2–4; ovate, dark green, 1–2 cm wide × 2–5 cm long, usually absent at flowering time
Flowers: 10–35; in a single rank, spiraled or secund; sepals and petals similar, elliptic; perianth white; lip oblong, with the apex rounded; central portion green with a clearly defined crisp apron; individual flower size ca. 4.0–7.5 mm
Habitat: dry to moist meadows, grassy roadsides, cemeteries, open sandy areas in woodlands, lawns, old fields
Flowering period: summer flowering; July through August and often into early September

Spiranthes lacera var. *gracilis*, the **southern slender ladies'-tresses,** is the more southerly of the two varieties and found in nearly every state in the central and eastern United States. The small, white, green-throated flowers are very distinctive and the simple spiral of the inflorescence quite eye-catching. Plants are often encountered in lawns, and local cemeteries are usually a good place to search as well. The differences between this variety and the more northerly var. *lacera* are not great, but the more northern of the two has lower flowers well spaced out on the inflorescence and the flowers appear to be much smaller because of the position of the sepals.

Spiranthes laciniata (Small) Ames
lace-lipped ladies'-tresses

Range: eastern Texas east through all of Florida and north to southern New Jersey; primarily on the Coastal Plain
Within the prairies and Great Plains region:
Louisiana, Texas: widely scattered and locally frequent in the southern wet prairies
Plant: terrestrial, 20–95 cm tall; densely pubescent with capitate hairs
Leaves: 3–5; lanceolate, 1.0–1.7 cm wide × 5–40 cm long
Flowers: 10–50; in a single rank, spiraled or secund; sepals and petals similar, elliptic; perianth white to ivory; lip oblong, with the apex undulate-lacerate, the central portion of the lip creamy yellow; individual flower size ca. 1 cm
Habitat: wet, grassy roadsides, ditches, swamps, and shallow open water
Flowering period: late May through July

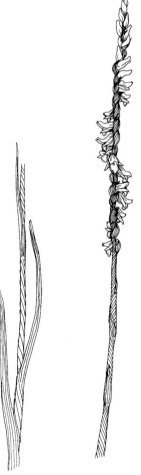

Spiranthes laciniata, the **lace-lipped ladies'-tresses**, is easily distinguished from *S. vernalis*, the **grass-leaved ladies'-tresses**, which it superficially resembles, by its ball-tipped hairs rather than the pointed articulate hairs found on *S. vernalis*. Where the two species are sympatric *S. laciniata* typically flowers later and in wetter habitats than *S. vernalis*. The tall, creamy white spikes may be a frequent sight along the wet roadside ditches and open savannas of the Gulf Coastal Plain. Plants originally identified as *S. laciniata* from Arkansas have been correctly identified as *S. vernalis* (Johnson, 2000, 2004).

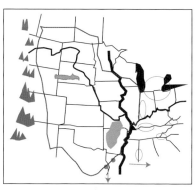

Spiranthes lucida (H.H. Eaton) Ames
shining ladies'-tresses

Range: Wisconsin east to Nova Scotia, south to Kansas, Alabama, and West Virginia
Within the prairies and Great Plains region:
Illinois, Iowa, Kansas, Missouri, Nebraska, Wisconsin: rare and local along calcareous riverbanks and fens
Plant: terrestrial, 10–38 cm tall
Leaves: 3–4; basal, elliptic-lanceolate, 0.5–1.5 cm wide × 3–12 cm long; the surface shiny; present at flowering time and long beyond
Flowers: 10–20; in an open spiral; nearly horizontal to nodding; sepals and petals similar, white, lanceolate; lateral sepals appressed to the petals and lip forming a tube; the lip oblong, bright yellow with the central portion yellow-orange or green, the margin crenulate; individual flower size 0.6–0.9 mm
Habitat: rocky riverbanks, seeps, fens; usually calcareous
Flowering period: May well into June

The **shining ladies'-tresses**, *Spiranthes lucida*, is unlike any other in the genus in North America. The broad, shining leaves and tubular yellow-lipped flowers set it apart. In the far northern portion of its range the **shining ladies'-tresses** is often accompanied by plants of *S. romanzoffiana*, the **hooded ladies'-tresses**, that flower in the summer or autumn. In the prairies and Great Plains *S. lucida* is found along calcareous river shores. *Spiranthes lucida* is historic in Kansas and was discovered in Iowa in 1987 (Witt, 2006).

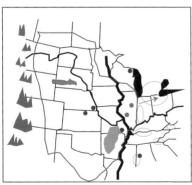

Spiranthes magnicamporum Sheviak
Great Plains ladies'-tresses

Range: Manitoba east to southern Ontario, south to New Mexico, Texas, Pennsylvania, and Georgia
Within the prairies and Great Plains region:
Manitoba; Colorado, Illinois, Iowa, Kansas, Louisiana, Minnesota, Missouri, New Mexico, Nebraska, North Dakota, Oklahoma, South Dakota, Wisconsin, Texas: widespread locales throughout the Great Plains and prairies
Plant: terrestrial, 10–63 cm tall
Leaves: 2–4; appearing basal or on the lower portion of the stem; linear-oblanceolate, up to 1.5 cm wide × 16.0 cm long; ascending to spreading; leaves usually absent or withering at anthesis
Flowers: 12–54; in a spike, tightly to loosely spiraled with 3–4 flowers per cycle; abruptly nodding from the base; lateral sepals wide-spreading and usually arched above the flower, petals linear-lanceolate; perianth white, ivory, or cream; lip ovate to oblong, the apex crenulate, the central portion of the lip usually yellow; individual flower size 0.4–1.2 cm
Habitat: wet to dry alkaline prairies, bluffs, and fens
Flowering period: September through November southward

The **Great Plains ladies'-tresses**, *Spiranthes magnicamporum*, a typical prairie species, is reaching the southeastern limit of its range in several widely scattered and nearly disjunct sites in the Southeast. The **Great Plains ladies'-tresses** is one of several species of *Spiranthes* that contribute gene flow to the **nodding ladies'-tresses**, *S. cernua*. Several of these eastern prairie sites contain mixed populations of both *S. magnicamporum* and *S. cernua*, so careful examination of the plants and, if absolutely necessary, the seeds is essential to determine the correct species. Seeds from the basic diploid species *S. magnicamporum*, *S. ochroleuca*, *S. odorata*, and *S. ovalis* are monoembryonic whereas those of *S. cernua*, usually a polyploid species, are polyembryonic. This type of examination goes beyond the scope of this field guide and requires more sophisticated techniques. A compound microscope and the skills to use it are required.

Spiranthes odorata (Nuttall) Lindley

fragrant ladies'-tresses

Range: eastern Texas north to Oklahoma and Arkansas, east to Florida, and north to (?)Delaware
Within the prairies and Great Plains region:
Arkansas, Louisiana, Oklahoma, Texas: widespread and locally common especially through several of the southern states
Plant: terrestrial or semi-aquatic, 20–110 cm tall, pubescent, stoloniferous
Leaves: 3–5; linear-oblanceolate, up to 4 cm wide × 52 cm long; rigidly ascending or spreading
Flowers: 10–30; in several tight ranks; sepals and petals similar, lanceolate; perianth white to ivory; lip oblong, tapering to the apex, the central portion of the lip creamy-yellow or green; the sepals extending forward; individual flower size 1.0–1.8 cm
Habitat: moist grassy roadsides, pine flatwoods, cypress swamps, wooded river floodplains
Flowering period: October into early December in the south

Spiranthes odorata, the **fragrant ladies'-tresses**, can be by far the largest of our native ladies'-tresses. Plants in wooded swamplands can be locally abundant and reach a full meter in height. The **fragrant ladies'-tresses** typically occurs in seasonally inundated sites and may flower while emerging from shallow water. The rather thick, broad leaves give the plant a distinctive habit. The very long, wide-spreading roots produce vegetative offshoots often 30 cm from the parent shoot, giving rise to extensive clonal colonies. Despite the typical size of *S. odorata*, there is a very definable ecotype that occupies mowed road shoulders and can often be no more than 15 cm tall. Hybrids with *Spiranthes ovalis* are known as *S.* ×*itchetuckneensis*.

Spiranthes ovalis Lindley var. *ovalis*

southern oval ladies'-tresses

Range: Arkansas south to eastern Texas, east to Florida
Globally Threatened
Within the prairies and Great Plains region:
Arkansas, Louisiana, Oklahoma, Texas: very rare and local with only a few populations in each state
Plant: terrestrial, 20–40 cm tall; pubescent
Leaves: 2–4; basal and on the lower half of the stem, oblanceolate, 0.5–1.5 cm wide × 3.0–15.0 cm long; present at flowering time
Flowers: 10–50; in 3 tight ranks; sepals and petals similar, lanceolate; perianth white; lip oblong, tapering to the apex with a delicate undulate margin, the sepals extending forward; individual flower size 5.5–7.0 mm; rostellum and viscidium present, therefore the plants are fully sexual
Habitat: rich, damp woodlands and floodplains
Flowering period: October into November

One of the most charming of all the ladies'-tresses, *Spiranthes ovalis*, the **southern oval ladies'-tresses**, is the only species with an exclusive woodland habitat. The pristine, small white flowers are usually carried in three distinctive, tight, vertical ranks. In the variety *ovalis* the flowers are sexually complete and therefore fertilization is effected by a pollinator. The flowers in variety *ovalis* are always fully expanded. This nominate variety is relatively rare and found only in a few states. Hybrids with *Spiranthes odorata* are known as *S.* ×*itchetuckneensis*.

Spiranthes ovalis Lindley var. *erostellata* Catling

northern oval ladies'-tresses

Range: Ontario south to Illinois and Arkansas, east to Ohio and western Pennsylvania, south to northern Florida
Within the prairies and Great Plains region:
Arkansas, Illinois, Iowa, Kansas, Louisiana, Missouri, Oklahoma, Texas: wide-ranging although scattered locales; apparently becoming more common, especially in disturbed woodland sites
Plant: terrestrial, 20–40 cm tall, pubescent
Leaves: 2–4; basal and on the lower half of the stem, oblanceolate, 0.5–1.5 cm wide × 3.0–15.0 cm long, present at flowering time
Flowers: 10–50; in 3 tight ranks; sepals and petals similar, lanceolate; perianth white; lip oblong, tapering to the apex with a delicate undulate margin; the sepals extending forward; individual flower size 3.5–5.0 mm; rostellum lacking, therefore the plants self-pollinating; flowers not always fully open
Habitat: rich, damp woodlands and floodplains
Flowering period: late September northward into November in the southern states

In most plants of *Spiranthes ovalis* var. *erostellata*, the **northern oval ladies'-tresses**, the flowers are never quite fully open, or in many individuals they are tiny cleistogamous flowers, and the ovaries are simultaneously swollen on each flower within the inflorescence. Those of *S. ovalis* var. *ovalis* have fully open flowers and the ovaries swell progressively. In a few sites in the southern states both varieties grow together as well as the **fragrant ladies'-tresses**, *S. odorata*. The hybrid between them, *S.* ×*itchetuckneensis*, is often also present. For more information on hybrids within the genus *Spiranthes* see page 206.

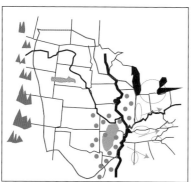

Spiranthes romanzoffiana Chamisso

hooded ladies'-tresses

Range: Alaska east to Newfoundland; south to California, northern New Mexico, Indiana, Pennsylvania; Northern Ireland, Great Britain
Within the prairies and Great Plains region:
Manitoba, Saskatchewan; Illinois, Iowa, Minnesota, Nebraska, New Mexico, North Dakota, South Dakota, Wisconsin: widespread throughout the northern Great Plains and prairies
Plant: terrestrial, 10–50 cm tall
Leaves: 3–6; basal and extending up the lower fourth of the stem; linear-lanceolate, up to 1.5 cm wide × 25.0 cm long, passing into a few slender bracts; present at flowering time
Flowers: 10–60; in a dense spiral; nearly horizontal to ascending; sepals and petals similar, creamy white to greenish-white to creamy-yellow, lanceolate, acuminate; lateral sepals appressed to the petals and lip forming an ascending hood; the lip oblong, panduriform, the broadened margin recurved and finely lacerate; individual flower size 0.9–1.2 cm
Habitat: rocky riverbanks, seeps, fens; usually calcareous
Flowering period: mid-July in the north and well into August southward

The **hooded ladies'-tresses**, *Spiranthes romanzoffiana*, is the most widespread *Spiranthes* found in northern North America. Starting to flower in midsummer in the far north, the species continues flowering until late August further south. The almond-scented flowers and arching hood are distinctive among our eastern *Spiranthes* and could not possibly be mistaken for any other species. In a few areas well northeast of the prairies and plains it hybridizes with *S. lacera* var. *lacera* to produce *Spiranthes* ×*simpsonii*.

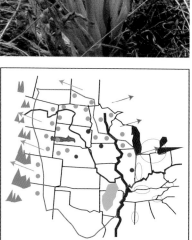

Spiranthes sylvatica P.M. Brown

woodland ladies'-tresses

Range: eastern Texas to Florida, north to Arkansas and along the Coastal Plain to southeastern North Carolina
Within the prairies and Great Plains region: Arkansas, Louisiana, Oklahoma, Texas: locally common in varied woodland habitats throughout the region; recent reports (Slaughter, pers. comm.) indicate that all the plants previously thought to be *Spiranthes praecox* in Arkansas are *S. sylvatica* and photographs from Oklahoma indicate the same
Plant: terrestrial, 25–75 cm tall, sparsely pubescent with capitate hairs
Leaves: 3–7; basal and on the lower third of the stem, linear-lanceolate, 0.8–1.57 cm wide × 10.0–35.0 cm long, present at flowering time
Flowers: 10–30; in a dense spike usually appearing as multiple ranks; sepals and petals similar, lanceolate; perianth creamy-green; lip ovate-oblong, rounded and broadened to the apex with a delicate undulate or ruffled margin with distinctive darker green veining, the sepals slightly spreading; individual flower size 1.0–1.7(2.2) cm
Habitat: shaded roadsides, open woodlands, and dry live oak hammocks; rarely in wetlands
Flowering period: April through early May

Spiranthes sylvatica, the **woodland ladies'-tresses**, is the most recent *Spiranthes* species described from North America (in Brown, 2001). Although the plants have been known for some time, sufficient evidence has only recently been available to satisfactorily separate this species from the giant ladies'-tresses, *Spiranthes praecox*. Although both typically have green-veined lips, all similarity ceases at that point. The **woodland ladies'-tresses** has been passed over for many years as a disappointing example of *S. praecox*. *Spiranthes sylvatica* is usually a plant of shaded and woodland habitats and the very distinctive, large, creamy-green flowers are unlike those of any other *Spiranthes*. It is most frequently seen along roadside hedgerows bordering woodlands, where the plants are tucked up into the border. There are many additional distinctive differences between the two species but flower size, shape, and color are the most noticeable. Plants have the habit of producing long, slender leaves in the autumn that winter over until new growth commences in the early spring. Recent fieldwork has confirmed extant sites for Texas, Arkansas, Georgia, South Carolina, and North Carolina.

winter leaves

Wild Orchids of the Prairies and Great Plains Region of North America

Spiranthes tuberosa Rafinesque
little ladies'-tresses

Range: Arkansas east to southern Michigan and Massachusetts, south to Florida and west to Texas
Within the prairies and Great Plains region:
Arkansas, Illinois, Kansas, Louisiana, Missouri, Oklahoma, Texas: a common summer-flowering species particularly in the southern states
Plant: terrestrial, 10–30 cm tall, glabrous
Leaves: 2–4; ovate, dark green, 1–2 cm wide × 2–5 cm long, absent at flowering time
Flowers: 10–35; in a single rank, spiraled or secund; sepals and petals similar, elliptic; perianth crystalline white; lip oblong, with the apex undulate-lacerate, exceeding the sepals; individual flower size ca. 3–4 mm
Habitat: grassy roadsides, cemeteries, open sandy areas in woodlands
Flowering period: highly variable from late June through early October

Spiranthes tuberosa, the **little ladies'-tresses**, is the only pure white *Spiranthes* to flower in early to mid-summer. One of its favorite habitats is old, dry cemeteries. The nomenclatural history of this plant is rather complex, and among the names applied to it are *S. beckii* and *S. grayi*. See Correll (1950) for a discussion. This species may be easily recognized by its pure white flowers, broad crisped lip, and absence of leaves at flowering time.

Spiranthes vernalis Engelmann & Gray
grass-leaved ladies'-tresses

Range: Nebraska south to Texas, east to Florida, and north to southern New Hampshire
Within the prairies and Great Plains region:
Arkansas, Illinois, Iowa, Kansas, Louisiana, Missouri, Nebraska, Oklahoma, South Dakota, Texas: the most common *Spiranthes* found in the southeastern United States
Plant: terrestrial, 10–85 cm tall, pubescent with sharp-pointed hairs
Leaves: 2–7; basal and on the lower third of the stem, linear-lanceolate, 1–2 cm wide × 5–25 cm long, present at flowering time
Flowers: 10–50; arranged in either a single or multiple ranks; sepals and petals similar, lanceolate; perianth typically creamy-white; lip ovate-oblong, rounded to the apex with a delicate undulate margin, usually a deeper creamy-yellow; the sepals wide-spreading; individual flower size 6–9 mm
Habitat: roadsides, meadows, prairies, cultivated lawns—just about anywhere that is sunny
Flowering period: April and May in the southern and central states, often persisting until late August at the northern limit of its range

Spiranthes vernalis, the **grass-leaved ladies'-tresses,** is perhaps our most variable *Spiranthes* in habit, although the flowers remain surprisingly consistent. Plants may vary greatly in size and vigor, as well as degree of spiraling, resulting in plants essentially secund to those that appear to be multiple-ranked. Plants are not consistent in habit from year to year. Color is somewhat variable from nearly pure white to cream with a contrasting, yellower lip; some individuals have two brown or orange spots on the lip. The most consistent diagnostic character is the presence in the inflorescence of copious articulate, pointed hairs that readily distinguish *S. vernalis* from other species.

extreme coloration found in alkali or high salt areas

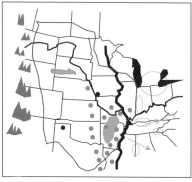

Hybrids:

Spiranthes ×*intermedia* Ames
intermediate hybrid ladies'-tresses
(*S. lacera* var. *gracilis* × *S. vernalis*)
  Rhodora 5: 262, pl. 47. 1903, type:
    Massachusetts

One of the first *Spiranthes* hybrids described, it only occurs in a few places, primarily in the Northeast, where both parents flower at the same time.

The presence of both parental species in the Midwest may lead to the discovery of this hybrid in the central prairies and Great Plains states.

Spiranthes ×*itchetuckneensis* P.M. Brown
Ichetucknee hybrid ladies'-tresses
(*S. odorata* × *S. ovalis*)
  North American Native Orchid Journal 5(4): 358–367.

This hybrid is frequent in the Deep South and could occur in the southern prairies region in wet woodlands where both parents grow together, especially in southern Arkansas.

Spiranthes ×*simpsonii* Catling & Sheviak
Simpson's hybrid ladies'-tresses
(*S. lacera* var. *lacera* × *S. romanzoffiana*)
  Lindleyana 8(2): 78–80. 1993, type: Ontario

Although the two parents often grow in proximity in the northern prairies and Great Plains region the hybrid is known from only a few collections in Ontario and northern New England and perhaps has been overlooked in other areas in the past.

Tipularia

Tipularia is a small genus of only two species known from the Himalayas and eastern United States. The species are very similar to each other and differ in the shape of the lip. The genus is characterized by having a series of tubers from which arise a single, an annual leaf that remain green throughout the winter, and then in the summer, well after the leaf withers, a leafless spike of flowers that has the sepals and petals all drawn to one side appears. A single species in North America is the **crane-fly orchid**, *Tipularia discolor*.

Tipularia discolor (Pursh) Nuttall
crane-fly orchis

forma *viridifolia* P.M. Brown—green-leaved form
North American Native Orchid Journal 6(4): 334–335. 2000, type: Florida

Range: eastern Texas northeastward to southern Michigan and east to southeastern Massachusetts, south to Florida
Within the prairies and Great Plains region:
Arkansas, Illinois, Louisiana, Missouri, Oklahoma, Texas: common throughout most of the Southeast
Plant: terrestrial, 25–60 cm tall
Leaves: 1; basal, ovate, dark green above with raised purple spots and dark purple beneath or, in the forma *viridifolia*, green on both sides; 6–7 cm wide × 8–10 cm, the long petiole ca. 5 cm; leaf withering in the spring and absent at flowering time
Flowers: 20–40; in a loose raceme; sepals and petals similar, oblanceolate; perianth greenish-yellow, tinged and mottled with pale purple; lip 3-lobed, the central lobe slender, blunt, with a few shallow teeth; the sepals and petals are asymmetrical and all drawn to one side; spur to 2.5–3.0 cm; individual flower size 2.0–3.0 cm × 3.0–3.5 cm not including spur
Habitat: deciduous and mixed woodlands
Flowering period: late June in the southern states and eventually August in the north

The **crane-fly orchis**, *Tipularia discolor*, is a very distinctive species most easily found during the winter months when not in flower and the single leaf is most apparent. The dark green leaf has a seersucker look with raised purple spots on the upper surface. If one turns the leaf over the satiny purple underside can be seen—hence the name *discolor* or two colors. Plants with no purple present, forma *viridifolia*, have been seen in scattered locations. The flower spike appears in midsummer and the coloring of the stem and flowers makes it most difficult to see in the woodlands, although the spike and flowers are not at all small. Spikes usually grow to 45–50+ cm tall.

Triphora

Consisting of about twenty species in North America, the West Indies, Mexico, and Central America, *Triphora* is a genus of small, delicate, fleshy plants, many of which may be largely mycotrophic. They all arise from swollen tuberoids and some produce very colorful, although small, flowers. Several species have flowers that do not fully open. In the prairies and Great Plains we have only a single species.

Triphora trianthophora (Swartz) Rydberg subsp. *trianthophora*

three birds orchid

forma *albidoflava* Keenan—white-flowered form
   Rhodora 94: 38–39. 1992, type: New Hampshire
forma *caerulea* P.M. Brown—blue-flowered form
   North American Native Orchid Journal 7(1): 94–95. 2001, type: Florida
forma *rossii* P.M. Brown—multi-colored form
   North American Native Orchid Journal 5(1): 5. 1999, type: Florida
Range: Texas north to Minnesota, east to Maine, south to Florida
Within the prairies and Great Plains region: Arkansas, Illinois, Iowa, Kansas, Louisiana, Missouri, Nebraska, Oklahoma, Texas, Wisconsin: rare and local throughout
Plant: terrestrial, 8–25 cm tall
Leaves: 2–8; broadly ovate-cordate, with smooth margins, dark green often with a purple cast or, in the forma *rossii*, the stem and leaves white, pink, and yellow; 1.0–1.5 cm × 0.2–1.5 cm
Flowers: 1–8 (12), nodding; from the axils of the upper leaves; sepals and petals similar, oblanceolate; perianth white to pink or, in the forma *albidoflava*, the perianth pure white and the crests yellow or, in the forma *caerulea*, the perianth lilac-blue; lip 3-lobed, the central lobe with the margin sinuate and 3 parallel green crests; individual flower size ca. 1–2 cm
Habitat: deciduous and mixed woodlands, often with partridge berry
Flowering period: varies from late July until mid-September

The **three birds orchid**, *Triphora trianthophora*, is the largest-flowered and showiest of the genus *Triphora*. The plants are quite elusive and appear for only a few days most years. The stunning little flowers open in midmorning and usually close by midafternoon, leaving only a few hours for the eager eye to observe them. Colonies are not at all consistent in their flowering habits from year to year and it often takes a great deal of persistence on the part of the observer to catch them in prime condition. Plants have a decided preference for live oak and beech woodlands; colonies are often widely scattered and few-flowered and may remain dormant for several years. *Triphora trianthophora* subsp. *mexicana* has occasionally been reported from Florida but no vouchers could be found. It is restricted to Mexico.

forma *albidoflava*

Bordering species

The majority of orchid species that approach the prairies and Great Plains region are those, especially in the north, of the hardwood and boreal forest. Those species would not find suitable habitat in the open grasslands although *Arethusa bulbosa* could conceivably inhabit a northern prairie fen along with *Calopogon tuberosus* and *Pogonia ophioglossoides*. Additional southern species may also border the central grasslands of southern North America but are normally found within the swamps and woodlands of the south. They might include *Ponthieva racemosa*, the shadow-witch orchid, and *Platythelys querceticola*, the low ground orchid.

Three species that could be considered valid bordering species are found in Texas and/or Louisiana and are typical of the Gulf Coastal Plain rather than the prairie regions. They should be carefully sought in the southern prairie regions of those states.

Calopogon barbatus
bearded grass-pink
Previously considered to range west throughout Louisiana, east Texas, and southern Arkansas, most of those plants have proven to be *Calopogon oklahomensis* (Goldman, 1995). A few sites remain for *C. barbatus* in western Louisiana.

Gymnadeniopsis integra
yellow fringeless orchis
Rarest of the rare, the yellow fringeless orchis is a plant of southern savannas but could possibly cross over to the prairie areas.

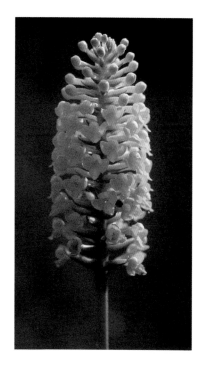

Pteroglossaspis ecristata
crestless plume orchis
A recent find in western Louisiana makes this a bordering species. Open grasslands and scrubby hedgerows are some of its preferred habitats.

Part 3

~

References and Resources

Checklist of the Wild Orchids of the North American Prairies and Great Plains Region

Aplectrum hyemale (Mühlenberg *ex* Willdenow) Nuttall
putty-root, Adam-and-Eve
    forma *pallidum* House—yellow-flowered form

Calopogon oklahomensis D.H. Goldman
Oklahoma grass-pink
    forma *albiflorus* P.M. Brown—white-flowered form

Calopogon tuberosus (Linnaeus) Britton, Sterns, & Poggenberg var. *tuberosus*
common grass-pink
    forma *albiflorus* Britton—white-flowered form

Calypso bulbosa (Linnaeus) Oakes var. *americana* (R. Brown) Luer
eastern fairy slipper
    forma *albiflora* P.M. Brown—white-flowered form
    forma *biflora* P.M. Brown—two-flowered form
    forma *rosea* P.M. Brown—pink-flowered form

Coeloglossum viride (Linnaeus) Hartman var. *virescens* (Mühlenberg) Luer
long-bracted green orchis

Corallorhiza maculata (Rafinesque) Rafinesque var. *maculata*
spotted coralroot
    forma *flavida* (Peck) Farwell—yellow-stemmed form
    forma *rubra* P.M. Brown—red-stemmed form
Corallorhiza maculata (Rafinesque) Rafinesque var. *occidentalis* (Lindley) Ames
western spotted coralroot
    forma *aurea* P.M. Brown—golden yellow/spotted form
    forma *immaculata* (Peck) Howell—yellow spotless form
    forma *intermedia* Farwell—brown-stemmed form
    forma *punicea* (Bartholomew) Weatherby & Adams—red-stemmed form

Corallorhiza odontorhiza (Willdenow) Nuttall var. *odontorhiza*
autumn coralroot
    forma *flavida* Wherry—yellow-flowered form

Corallorhiza odontorhiza (Willdenow) Nuttall var. *pringlei* (Greenman) Freudenstein
Pringle's autumn coralroot

Corallorhiza striata Lindley var. *striata*
striped coralroot
    forma *eburnea* P.M. Brown—yellow/white form
    forma *fulva* Fernald—dusky tan-colored form
Corallorhiza striata Lindley var. *vreelandii* (Rydberg) L.O. Williams
Vreeland's striped coralroot
    forma *flavida* (Todsen & Todsen) P.M. Brown—yellow/white form

Corallorhiza trifida Chatelain
early coralroot
    forma *verna* (Nuttall) P.M. Brown—yellow-stemmed/white-lipped form

Corallorhiza wisteriana Conrad
Wister's coralroot
    forma *albolabia* P.M. Brown—white-lipped form
    forma *cooperi* P.M. Brown—cranberry-pink colored form
    forma *rubra* P.M. Brown—red-stemmed form

Cypripedium acaule Aiton
pink lady's-slipper, moccasin flower
    forma *albiflorum* Rand & Redfield—white-flowered form
    forma *biflorum* P.M. Brown—two-flowered form
    forma *lancifolium* House—narrow-leaved form

Cypripedium candidum Mühlenberg *ex* Willdenow
small white lady's-slipper

Cypripedium kentuckiense C.F. Reed
ivory-lipped lady's-slipper, Kentucky lady's-slipper
    forma *pricei* P.M. Brown—white-flowered form
    forma *summersii* P.M. Brown—concolorous yellow-flowered form

Cypripedium parviflorum Salisbury var. *parviflorum*
southern small yellow lady's-slipper
    forma *albolabium* Magrath & Norman—white-lipped form
Cypripedium parviflorum Salisbury var. *makasin* (Farwell) Sheviak
northern small yellow lady's-slipper
Cypripedium parviflorum Salisbury var. *pubescens* (Willdenow) Knight
large yellow lady's-slipper

Cypripedium reginae Walter
showy lady's-slipper
    forma *albolabium* Fernald & Schubert—white-flowered form
Hybrids:
Cypripedium ×*andrewsii* Fuller nm *andrewsii*
Andrews' hybrid lady's-slipper
Cypripedium ×*andrewsii* Fuller nm *favillianum* (Curtis) Boivin
Faville's hybrid lady's-slipper
Cypripedium ×*andrewsii* nm *landonii* (Garay) Boivin
Landon's hybrid lady's-slipper

Epipactis gigantea Douglas *ex* Hooker
stream orchid, chatterbox
    forma *citrina* P.M. Brown—yellow-flowered form
    forma *rubrifolia* P.M. Brown—red-leaved form

Epipactis helleborine (Linnaeus) Cranz
broad-leaved helleborine*
    forma *alba* (Webster) Boivin—white-flowered form
    forma *luteola* P.M. Brown—yellow-flowered form
    forma *monotropoides* (Mousley) Scoggin—albino form
    forma *variegata* (Webster) Boivin—variegated form
    forma *viridens* A. Gray—green-flowered form

Galearis spectabilis (Linnaeus) Rafinesque
showy orchis
    forma *gordinierii* (House) Whiting & Catling—white-flowered form
    forma *willeyi* (Seymour) P.M. Brown—pink-flowered form

Goodyera oblongifolia Rafinesque
giant rattlesnake orchis
    forma *reticulata* (Boivin) P.M. Brown—reticulated leaved form

Goodyera pubescens (Willdenow) R. Brown
downy rattlesnake orchis

Goodyera repens (Linnaeus) R. Brown
lesser rattlesnake orchis
    forma *ophioides* (Fernald) P.M. Brown—white-veined leaved form

Gymnadeniopsis clavellata (Michaux) Rydberg var. *clavellata*
little club-spur orchis
    forma *slaughteri* (P.M. Brown) P.M. Brown—white-flowered form
    forma *wrightii* (Olive) P.M. Brown—spurless form

Gymnadeniopsis nivea (Nuttall) Rydberg
snowy orchis

Habenaria repens Nuttall
water-spider orchis

Hexalectris spicata (Walter) Barnhardt var. *spicata*
crested coralroot
    forma *albolabia* P.M. Brown—white-lipped form
    forma *lutea* P.M. Brown—yellow-flowered form
    forma *wilderi* P.M. Brown—albino form

Isotria medeoloides (Pursh) Rafinesque
small whorled pogonia

Isotria verticillata (Mühlenberg *ex* Willdenow) Rafinesque
large whorled pogonia

Liparis liliifolia (Linnaeus) Richard *ex* Lindley
lily-leaved twayblade
    forma *viridiflora* Wadmond—green-flowered form

Liparis loeselii (Linnaeus) Richard
Loesel's twayblade, fen orchis
Hybrid:
Liparis ×*jonesii* S. Bentley
Jones' hybrid twayblade

Listera australis Lindley
southern twayblade
    forma *scottii* P.M. Brown—many-leaved form
    forma *trifolia* P.M. Brown—three-leaved form
    forma *viridis* P.M. Brown—green-flowered form

Listera convallarioides (Swartz) Nuttall
broad-lipped twayblade
    forma *trifolia* P.M. Brown—three-leaved form

Malaxis unifolia Michaux
green adder's-mouth
    forma *bifolia* Mousley—two-leaved form
    forma *variegata* Mousley—variegated-leaf form

Piperia unalascensis (Sprengel) Rydberg
Alaskan piperia
    forma *olympica* P.M. Brown—dwarf montane form

Platanthera aquilonis Sheviak
northern green bog orchis
    forma *alba* (Light) P.M. Brown—albino form

Platanthera blephariglottis (Willdenow) Lindley
northern white fringed orchis
    forma *holopetala* (Lindley) P.M. Brown—entire-lip form

Platanthera ciliaris (Linnaeus) Lindley
orange fringed orchis

Platanthera cristata (Michaux) Lindley
orange crested orchis
    forma *straminea* P.M. Brown—pale yellow-flowered form

Platanthera dilatata (Pursh) Lindley var. *dilatata*
tall white northern bog orchis
Platanthera dilatata (Pursh) Lindley var. *albiflora* (Chamisso) Ledebour
bog candles

Platanthera flava (Linnaeus) Lindley var. *flava*
southern tubercled orchis
Platanthera flava (Linnaeus) Lindley var. *herbiola* (R. Brown) Luer
northern tubercled orchis
    forma *lutea* (Boivin) Whiting & Catling—yellow-flowered form

Platanthera hookeri (Torrey) Lindley
Hooker's orchis
    forma *abbreviata* (Fernald) P.M. Brown—dwarfed form
    forma *oblongifolia* (J.A. Paine) P.M. Brown—narrow-leaved form

Platanthera huronensis (Nuttall) Lindley
green bog orchis

Platanthera lacera (Michaux) G. Don
green fringed orchis; ragged orchis

Platanthera leucophaea (Nuttall) Lindley
eastern prairie fringed orchis

Platanthera orbiculata (Pursh) Lindley
pad-leaved orchis
    forma *lehorsii* (Fernald) P.M. Brown—dwarfed form
    forma *longifolia* (Clute) P.M. Brown—narrow-leaved form
    forma *pauciflora* (Jennings) P.M. Brown—few-flowered form
    forma *trifolia* (Mousley) P.M. Brown—three-leaved form

Platanthera peramoena (A. Gray) A. Gray
purple fringeless orchis
    forma *doddsiae* P.M. Brown—white-flowered form

Platanthera praeclara Sheviak & Bowles
western prairie fringed orchis

Platanthera psycodes (Linnaeus) Lindley
small purple fringed orchis
    forma *albiflora* (R. Hoffman) Whiting & Catling—white-flowered form
    forma *ecalcarata* (Bryan) P.M. Brown—spurless form
    forma *fernaldii* (Rousseau & Rouleau) P.M. Brown—dwarf form
    forma *rosea* P.M. Brown—pink-flowered form
    forma *varians* (Bryan) P.M. Brown—entire-lip form

Platanthera stricta Lindley
slender bog orchis
Hybrids:
Platanthera ×andrewsii (Niles) Luer
Andrews' hybrid fringed orchis
Plantanthera ×channellii Folsom
Channell's fringed orchis

Pogonia ophioglossoides (Linnaeus) Ker-Gawler
rose pogonia, snakemouth orchid
    forma *albiflora* Rand & Redfield—white-flowered form
    forma *brachypogon* (Fernald) P.M. Brown—short-bearded form

Spiranthes casei Catling & Cruise var. *casei*
Case's ladies'-tresses

Spiranthes cernua (Linnaeus) L.C. Richard
nodding ladies'-tresses

Spiranthes diluvialis Sheviak
Ute ladies'-tresses

Spiranthes lacera Rafinesque var. *lacera*
northern slender ladies'-tresses
Spiranthes lacera Rafinesque var. *gracilis* (Bigelow) Luer
southern slender ladies'-tresses

Spiranthes laciniata (Small) Ames
lace-lipped ladies'-tresses

Spiranthes lucida (H.H. Eaton) Ames
shining ladies'-tresses

Spiranthes magnicamporum Sheviak
Great Plains ladies'-tresses

Spiranthes odorata Lindley
fragrant ladies'-tresses

Spiranthes ovalis Lindley var. *ovalis*
southern oval ladies'-tresses
Spiranthes ovalis Lindley var. *erostellata* Catling
northern oval ladies'-tresses

Spiranthes romanzoffiana Chamisso
hooded ladies'-tresses

Spiranthes sylvatica P.M. Brown
woodland ladies'-tresses

Spiranthes tuberosa Rafinesque
little ladies'-tresses

Spiranthes vernalis Engelmann & Gray
grass-leaved ladies'-tresses
Hybrids:
Spiranthes ×*intermedia* Ames
intermediate hybrid ladies'-tresses
Spiranthes ×*itchetuckneensis* P.M. Brown
Ichetucknee hybrid ladies'-tresses
Spiranthes ×*simpsonii* Catling & Sheviak
Simpson's hybrid ladies'-tresses

Tipularia discolor (Pursh) Nuttall
crane-fly orchis
    forma *viridifolia* P.M. Brown—green-leaved form

Triphora trianthophora (Swartz) Rydberg subsp. *trianthophora*
three birds orchid
    forma *albidoflava* Keenan—white-flowered form
    forma *caerulea* P.M. Brown—blue-flowered form
    forma *rossii* P.M. Brown—multi-colored form

Provincial and State Lists

H = historical

Saskatchewan

11 species and varieties found in the prairie and plains regions, of 29 species and varieties found in the province
Corallorhiza maculata var. *maculata*, spotted coralroot
Corallorhiza maculata var. *occidentalis*, western spotted coralroot
Corallorhiza striata var. *striata*, striped coralroot
Corallorhiza trifida, early coralroot
Cypripedium candidum, small white lady's-slipper H
Cypripedium parviflorum var. *makasin*, northern small yellow lady's-slipper
Cypripedium parviflorum var. *pubescens*, large yellow lady's-slipper
Cypripedium reginae, showy lady's-slipper
Liparis loeselii, Loesel's twayblade
Platanthera aquilonis, northern green bog orchis
Spiranthes romanzoffiana, hooded slender ladies'-tresses

Manitoba

23 species and varieties found in the prairies and plains regions, of 38 species and varieties found in the province
Corallorhiza maculata var. *maculata*, spotted coralroot
Corallorhiza maculata var. *occidentalis*, western spotted coralroot
Corallorhiza striata var. *striata*, striped coralroot
Corallorhiza trifida, early coralroot
Cypripedium acaule, pink lady's-slipper
Cypripedium candidum, small white lady's-slipper
Cypripedium parviflorum var. *makasin*, northern small yellow lady's-slipper
Cypripedium parviflorum var. *pubescens*, large yellow lady's-slipper
Cypripedium reginae, showy lady's-slipper
Liparis loeselii, Loesel's twayblade
Malaxis unifolia, green adder's-mouth
Platanthera aquilonis, northern green bog orchis
Platanthera dilatata var. *dilatata*, tall white northern bog orchis

Platanthera hookeri, Hooker's orchis
Platanthera huronensis, green bog orchis
Platanthera lacera, green fringed orchis
Platanthera orbiculata, pad-leaved orchis
Platanthera praeclara, western prairie fringed orchis
Platanthera psycodes, small purple fringed orchis
Pogonia ophioglossoides, rose pogonia
Spiranthes lacera var. *lacera*, northern slender ladies'-tresses
Spiranthes magnicamporum, Great Plains ladies'-tresses
Spiranthes romanzoffiana, hooded ladies'-tresses

Arkansas

39 species and varieties found in the state
Aplectrum hyemale, putty-root
Calopogon oklahomensis, Oklahoma grass-pink
Calopogon tuberosus, common grass-pink
Corallorhiza odontorhiza var. *odontorhiza*, autumn coralroot
Corallorhiza wisteriana, Wister's coralroot
Cypripedium kentuckiense, ivory-lipped lady's-slipper
Cypripedium parviflorum var. *parviflorum*, southern small yellow lady's-slipper
Cypripedium parviflorum var. *pubescens*, large yellow lady's-slipper
Cypripedium reginae, showy lady's-slipper
Epipactis helleborine, broad-leaved helleborine*
Galearis spectabilis, showy orchis
Goodyera pubescens, downy rattlesnake orchis
Gymnadeniopsis clavellata, little club-spur orchis
Gymnadeniopsis nivea, snowy orchis H
Habenaria repens, water-spider orchis
Hexalectris spicata, crested coralroot
Isotria verticillata, large whorled pogonia
Liparis liliifolia, lily-leaved twayblade
Liparis loeselii, Loesel's twayblade
Listera australis, southern twayblade
Malaxis unifolia, green adder's-mouth
Platanthera ciliaris, orange fringed orchis
Platanthera cristata, orange crested orchis
Platanthera flava var. *flava*, southern tubercled orchis
Platanthera flava var. *herbiola*, northern tubercled orchis
Platanthera lacera, green fringed orchis
Platanthera peramoena, purple fringeless orchis
Pogonia ophioglossoides, rose pogonia
Spiranthes cernua, nodding ladies'-tresses

Spiranthes lacera var. *lacera*, northern slender ladies'-tresses
Spiranthes lacera var. *gracilis*, southern slender ladies'-tresses
Spiranthes odorata, fragrant ladies'-tresses
Spiranthes ovalis var. *ovalis*, southern oval ladies'-tresses
Spiranthes ovalis var. *erostellata*, northern oval ladies'-tresses
Spiranthes sylvatica, woodland ladies'-tresses
Spiranthes tuberosa, little ladies'-tresses
Spiranthes vernalis, grass-leaved ladies'-tresses
Tipularia discolor, crane-fly orchis
Triphora trianthophora, three birds orchid

Colorado

5 species and varieties are found in the plains and prairies regions, of 28 species and varieties found in the state
Corallorhiza trifida, early coralroot
Corallorhiza wisteriana, Wister's coralroot
Cypripedium parviflorum var. *pubescens*, large yellow lady's- slipper
Goodyera oblongifolia, giant rattlesnake orchis
Spiranthes diluvialis, Ute ladies'-tresses

Illinois

50 species and varieties found in the state
Aplectrum hyemale, putty-root
Calopogon oklahomensis, Oklahoma grass-pink H
Calopogon tuberosus, common grass-pink
Coeloglossum viride var. *virescens*, long-bracted green orchis
Corallorhiza maculata var. *maculata*, spotted coralroot
Corallorhiza maculata var. *occidentalis*, western spotted coralroot
Corallorhiza odontorhiza var. *odontorhiza*, autumn coralroot
Corallorhiza trifida, early coralroot
Corallorhiza wisteriana, Wister's coralroot
Cypripedium candidum, small white lady's-slipper
Cypripedium parviflorum var. *makasin*, northern small yellow lady's-slipper
Cypripedium parviflorum var. *parviflorum*, southern small yellow lady's-slipper
Cypripedium parviflorum var. *pubescens*, large yellow lady's-slipper
Cypripedium reginae, showy lady's-slipper
Epipactis helleborine, broad-leaved helleborine*
Galearis spectabilis, showy orchis
Goodyera pubescens, downy rattlesnake orchis
Gymnadeniopsis clavellata, little club-spur orchis
Hexalectris spicata, crested coralroot

Isotria medeoloides, small whorled pogonia
Isotria verticillata, large whorled pogonia
Liparis liliifolia, lily-leaved twayblade
Liparis loeselii, Loesel's twayblade
Listera australis, southern twayblade
Malaxis brachypoda, white adder's-mouth
Malaxis unifolia, green adder's-mouth
Platanthera aquilonis, northern green bog orchis
Platanthera blephariglottis, northern white fringed orchis H
Platanthera ciliaris, orange fringed orchis
Platanthera dilatata var. *dilatata*, tall white northern bog orchis
Platanthera flava var. *flava*, southern tubercled orchis
Platanthera flava var. *herbiola*, northern tubercled orchis
Platanthera hookeri, Hooker's orchis
Platanthera lacera, green fringed orchis
Platanthera leucophaea, eastern prairie fringed orchis
Platanthera orbiculata, pad-leaved orchis
Platanthera peramoena, purple fringeless orchis
Platanthera psycodes, small purple fringed orchis
Pogonia ophioglossoides, rose pogonia
Spiranthes cernua, nodding ladies'-tresses
Spiranthes lacera var. *lacera*, northern slender ladies'-tresses
Spiranthes lacera var. *gracilis*, southern slender ladies'-tresses
Spiranthes lucida, shining ladies'-tresses
Spiranthes magnicamporum, Great Plains ladies'-tresses
Spiranthes ovalis var. *erostellata*, northern oval ladies'-tresses
Spiranthes romanzoffiana, hooded ladies'-tresses
Spiranthes tuberosa, little ladies'-tresses
Spiranthes vernalis, grass-leaved ladies'-tresses
Tipularia discolor, crane-fly orchis
Triphora trianthophora, three birds orchid

Iowa

32 species and varieties
Aplectrum hyemale, putty-root
Calopogon oklahomensis, Oklahoma grass-pink
Calopogon tuberosus, common grass-pink
Coeloglossum viride var. *virescens*, long-bracted green orchis
Corallorhiza maculata var. *maculata*, spotted coralroot
Corallorhiza odontorhiza var. *odontorhiza*, autumn coralroot
Corallorhiza odontorhiza var. *pringlei*, Pringle's autumn coralroot
Cypripedium candidum, small white lady's-slipper
Cypripedium parviflorum var. *makasin*, northern small yellow lady's-slipper

Cypripedium parviflorum var. *pubescens*, large yellow lady's-slipper
Cypripedium reginae, showy lady's-slipper
Galearis spectabilis, showy orchis
Goodyera pubescens, downy rattlesnake orchis
Gymnadeniopsis clavellata, little club-spur orchis
Liparis liliifolia, lily-leaved twayblade
Liparis loeselii, Loesel's twayblade
Malaxis unifolia, green adder's-mouth
Platanthera aquilonis, northern green bog orchis
Platanthera flava var. *herbiola*, northern tubercled orchis
Platanthera hookeri, Hooker's orchis
Platanthera leucophaea, eastern prairie fringed orchis
Platanthera praeclara, western prairie fringed orchis
Platanthera psycodes, small purple fringed orchis
Spiranthes cernua, nodding ladies'-tresses
Spiranthes lacera var. *lacera*, northern slender ladies'-tresses
Spiranthes lacera var. *gracilis*, southern slender ladies'-tresses
Spiranthes lucida, shining ladies'-tresses
Spiranthes magnicamporum, Great Plains ladies'-tresses
Spiranthes ovalis var. *erostellata*, northern oval ladies'-tresses
Spiranthes romanzoffiana, hooded ladies'-tresses
Spiranthes vernalis, grass-leaved ladies'-tresses
Triphora trianthophora, three birds orchid

Kansas

19 species and varieties
Aplectrum hyemale, putty-root
Calopogon oklahomensis, Oklahoma grass-pink
Corallorhiza odontorhiza var. *odontorhiza*, autumn coralroot
Corallorhiza wisteriana, Wister's coralroot
Cypripedium parviflorum var. *parviflorum*, southern small yellow lady's-slipper
Galearis spectabilis, showy orchis
Hexalectris spicata, crested coralroot
Liparis loeselii, Loesel's twayblade H
Malaxis unifolia, green adder's-mouth
Platanthera lacera, green fringed orchis
Platanthera praeclara, western prairie fringed orchis
Spiranthes cernua, nodding ladies'-tresses
Spiranthes lacera var. *lacera*, northern slender ladies'-tresses
Spiranthes lacera var. *gracilis*, southern slender ladies'-tresses
Spiranthes lucida, shining ladies'-tresses
Spiranthes magnicamporum, Great Plains ladies'-tresses

Spiranthes ovalis var. *erostellata*, northern oval ladies'-tresses
Spiranthes vernalis, grass-leaved ladies'-tresses
Triphora trianthophora, three birds orchid

Louisiana

27 species and varieties found in the prairies and plains regions, of 35 species and varieties found in the state
Calopogon oklahomensis, Oklahoma grass-pink
Calopogon tuberosus, common grass-pink
Corallorhiza odontorhiza var. *odontorhiza*, autumn coralroot
Corallorhiza wisteriana, Wister's coralroot
Gymnadeniopsis clavellata, little club-spur orchis
Gymnadeniopsis nivea, snowy orchis
Habenaria repens, water-spider orchis
Hexalectris spicata, crested coralroot
Isotria verticillata, large whorled pogonia
Listera australis, southern twayblade
Malaxis unifolia, green adder's-mouth
Platanthera ciliaris, orange fringed orchis
Platanthera cristata, orange crested orchis
Platanthera flava var. *flava*, southern tubercled orchis
Platanthera leucophaea, eastern prairie fringed orchis H
Pogonia ophioglossoides, rose pogonia
Spiranthes cernua, nodding ladies'-tresses
Spiranthes lacera var. *gracilis*, southern slender ladies'-tresses
Spiranthes magnicamporum, Great Plains ladies'-tresses
Spiranthes odorata, fragrant ladies'-tresses
Spiranthes ovalis var. *ovalis*, southern oval ladies'-tresses
Spiranthes ovalis var. *erostellata*, northern oval ladies'-tresses
Spiranthes sylvatica, woodland ladies'-tresses
Spiranthes tuberosa, little ladies'-tresses
Spiranthes vernalis, grass-leaved ladies'-tresses
Tipularia discolor, crane-fly orchis
Triphora trianthophora, three birds orchid

Minnesota

30 species and varieties found in the prairies and plains regions, of 47 species and varieties found in the state
Aplectrum hyemale, putty-root
Calopogon oklahomensis, Oklahoma grass-pink H
Calopogon tuberosus, common grass-pink

Corallorhiza maculata var. *maculata*, spotted coralroot
Corallorhiza maculata var. *occidentalis*, western spotted coralroot
Corallorhiza odontorhiza var. *odontorhiza*, autumn coralroot
Corallorhiza trifida, early coralroot
Cypripedium acaule, pink lady's-slipper
Cypripedium candidum, small white lady's-slipper
Cypripedium parviflorum var. *makasin*, northern small yellow lady's-slipper
Cypripedium parviflorum var. *pubescens*, large yellow lady's-slipper
Cypripedium reginae, showy lady's-slipper
Epipactis helleborine, broad-leaved helleborine*
Galearis spectabilis, showy orchis
Platanthera blephariglottis, northern white fringed orchis
Platanthera dilatata var. *dilatata*, tall white northern bog orchis
Platanthera flava var. *herbiola*, northern tubercled orchis
Platanthera hookeri, Hooker's orchis
Platanthera huronensis, green bog orchis
Platanthera lacera, green fringed orchis
Platanthera orbiculata, pad-leaved orchis
Platanthera praeclara, western prairie fringed orchis
Platanthera psycodes, small purple fringed orchis
Pogonia ophioglossoides, rose pogonia
Spiranthes cernua, nodding ladies'-tresses
Spiranthes lacera var. *lacera*, northern slender ladies'-tresses
Spiranthes lacera var. *gracilis*, southern slender ladies'-tresses
Spiranthes lucida, shining ladies'-tresses
Spiranthes magnicamporum, Great Plains ladies'-tresses
Spiranthes romanzoffiana, hooded ladies'-tresses

Missouri

39 species and varieties
Aplectrum hyemale, putty-root
Calopogon oklahomensis, Oklahoma grass-pink
Calopogon tuberosus, common grass-pink
Coeloglossum viride var. *virescens*, long-bracted green orchis
Corallorhiza odontorhiza var. *odontorhiza*, autumn coralroot
Corallorhiza wisteriana, Wister's coralroot
Cypripedium candidum, small white lady's-slipper
Cypripedium parviflorum var. *parviflorum*, southern small yellow lady's-slipper
Cypripedium parviflorum var. *pubescens*, large yellow lady's-slipper
Cypripedium reginae, showy lady's-slipper
Epipactis helleborine, broad-leaved helleborine*
Galearis spectabilis, showy orchis

Goodyera pubescens, downy rattlesnake orchis
Gymnadeniopsis clavellata, little club-spur orchis
Hexalectris spicata, crested coralroot
Isotria medeoloides, small whorled pogonia
Isotria verticillata, large whorled pogonia
Liparis liliifolia, lily-leaved twayblade
Liparis loeselii, Loesel's twayblade
Malaxis unifolia, green adder's-mouth
Platanthera ciliaris, orange fringed orchis
Platanthera flava var. *flava,* southern tubercled orchis
Platanthera flava var. *herbiola,* northern tubercled orchis
Platanthera lacera, green fringed orchis
Platanthera leucophaea, eastern prairie fringed orchis
Platanthera peramoena, purple fringeless orchis
Platanthera praeclara, western prairie fringed orchis
Platanthera psycodes, small purple fringed orchis
Pogonia ophioglossoides, rose pogonia
Spiranthes cernua, nodding ladies'-tresses
Spiranthes lacera var. *lacera,* northern slender ladies'-tresses
Spiranthes lacera var. *gracilis,* southern slender ladies'-tresses
Spiranthes lucida, shining ladies'-tresses
Spiranthes magnicamporum, Great Plains ladies'-tresses
Spiranthes ovalis var. *erostellata,* northern oval ladies'-tresses
Spiranthes tuberosa, little ladies'-tresses
Spiranthes vernalis, grass-leaved ladies'-tresses
Tipularia discolor, crane-fly orchis
Triphora trianthophora, three birds orchid

Montana

7 species and varieties found in the prairies and plains regions, of 34 species and varieties found in the state
Corallorhiza trifida, early coralroot
Cypripedium parviflorum var. *makasin,* northern small yellow lady's-slipper
Liparis loeselii, Loesel's twayblade
Platanthera aquilonis, northern green bog orchis
Platanthera dilatata var. *dilatata,* tall white northern bog orchis
Platanthera orbiculata, pad-leaved fringed orchis
Spiranthes romanzoffiana, hooded ladies'-tresses

Nebraska

20 species and varieties
Coeloglossum viride var. *virescens,* long-bracted green orchis

Corallorhiza maculata var. *occidentalis*, western spotted coralroot
Corallorhiza odontorhiza var. *odontorhiza*, autumn coralroot
Corallorhiza striata var. *striata*, striped coralroot
Corallorhiza wisteriana, Wister's coralroot
Cypripedium candidum, small white lady's-slipper
Cypripedium parviflorum var. *parviflorum*, southern small yellow lady's-slipper
Galearis spectabilis, showy orchis
Liparis loeselii, Loesel's twayblade
Platanthera aquilonis, northern green bog orchis
Platanthera dilatata var. *dilatata*, tall white northern bog orchis
Platanthera leucophaea, eastern prairie fringed orchis
Platanthera praeclara, western prairie fringed orchis
Spiranthes cernua, nodding ladies'-tresses
Spiranthes diluvialis, Ute ladies'-tresses
Spiranthes lucida, shining ladies'-tresses
Spiranthes magnicamporum, Great Plains ladies'-tresses
Spiranthes romanzoffiana, hooded ladies'-tresses
Spiranthes vernalis, grass-leaved ladies'-tresses
Triphora trianthophora, three birds orchid

New Mexico

8 species and varieties found in the prairies and plains regions, of 27 species and varieties found in the state
Corallorhiza odontorhiza var. *odontorhiza*, autumn coralroot
Corallorhiza trifida, early coralroot
Corallorhiza wisteriana, Wister's coralroot
Cypripedium parviflorum var. *pubescens*, large yellow lady's-slipper
Hexalectris spicata, crested coralroot
Piperia unalascensis, Alaskan piperia
Spiranthes magnicamporum, Great Plains ladies'-tresses
Spiranthes romanzoffiana, hooded ladies'-tresses

North Dakota

14 species and varieties
Coeloglossum viride var. *virescens*, long-bracted green orchis
Corallorhiza maculata var. *occidentalis*, western spotted coralroot
Corallorhiza striata var. *striata*, striped coralroot
Corallorhiza trifida, early coralroot
Cypripedium candidum, small white lady's-slipper
Cypripedium parviflorum var. *makasin*, northern small yellow lady's-slipper
Cypripedium parviflorum var. *pubescens*, large yellow lady's-slipper

Cypripedium reginae, showy lady's-slipper
Goodyera oblongifolia, giant rattlesnake orchis
Platanthera aquilonis, northern green bog orchis
Platanthera praeclara, western prairie fringed orchis
Pogonia ophioglossoides, rose pogonia
Spiranthes magnicamporum, Great Plains ladies'-tresses
Spiranthes romanzoffiana, hooded ladies'-tresses

Oklahoma

34 species and varieties
Aplectrum hyemale, putty-root
Calopogon oklahomensis, Oklahoma grass-pink
Calopogon tuberosus, common grass-pink
Corallorhiza odontorhiza var. *odontorhiza*, autumn coralroot
Corallorhiza wisteriana, Wister's coralroot
Cypripedium kentuckiense, ivory-lipped lady's-slipper
Cypripedium parviflorum var. *parviflorum*, southern small yellow lady's-slipper
Epipactis gigantea, stream orchid
Galearis spectabilis, showy orchis
Goodyera pubescens, downy rattlesnake orchis
Gymnadeniopsis clavellata, little club-spur orchis
Habenaria repens, water-spider orchis
Hexalectris spicata, crested coralroot
Isotria verticillata, large whorled pogonia
Liparis liliifolia, lily-leaved twayblade
Listera australis, southern twayblade
Malaxis unifolia, green adder's-mouth
Platanthera ciliaris, orange fringed orchis
Platanthera flava var. *flava*, southern tubercled orchis
Platanthera lacera, green fringed orchis
Platanthera leucophaea, eastern prairie fringed orchis
Platanthera praeclara, western prairie fringed orchis
Pogonia ophioglossoides, rose pogonia
Spiranthes cernua, nodding ladies'-tresses
Spiranthes lacera var. *gracilis*, southern slender ladies'-tresses
Spiranthes magnicamporum, Great Plains ladies'-tresses
Spiranthes odorata, fragrant ladies'-tresses
Spiranthes ovalis var. *ovalis*, southern oval ladies'-tresses
Spiranthes ovalis var. *erostellata*, northern oval ladies'-tresses
Spiranthes sylvatica, woodland ladies'-tresses
Spiranthes tuberosa, little ladies'-tresses
Spiranthes vernalis, grass-leaved ladies'-tresses

Tipularia discolor, crane-fly orchis
Triphora trianthophora, three birds orchid

South Dakota

25 species and varieties
Calypso bulbosa var. *americana*, eastern fairy slipper
Coeloglossum viride var. *virescens*, long-bracted green orchis
Corallorhiza maculata var. *occidentalis*, western spotted coralroot
Corallorhiza odontorhiza var. *odontorhiza*, autumn coralroot
Corallorhiza striata var. *striata*, striped coralroot
Corallorhiza striata var. *vreelandii*, Vreeland's striped coralroot
Corallorhiza trifida, early coralroot
Corallorhiza wisteriana, Wister's coralroot
Cypripedium candidum, small white lady's-slipper
Cypripedium parviflorum var. *pubescens*, small yellow lady's-slipper
Epipactis gigantea, stream orchid
Goodyera oblongifolia, giant rattlesnake orchis
Goodyera repens, lesser rattlesnake orchis
Listera convallarioides, broad-lipped twayblade
Piperia unalascensis, Alaskan piperia
Platanthera aquilonis, northern green bog orchis
Platanthera dilatata var. *dilatata*, tall white northern bog orchis
Platanthera dilatata var. *albiflora*, bog candles H
Platanthera huronensis, green bog orchis
Platanthera orbiculata, pad-leaved orchis
Platanthera praeclara, western prairie fringed orchis
Platanthera stricta, slender bog orchis H
Spiranthes cernua, nodding ladies'-tresses
Spiranthes magnicamporum, Great Plains ladies'-tresses
Spiranthes romanzoffiana, hooded ladies'-tresses

Texas

28 species and varieties, including 1 report, found in the prairies and plains regions, of 50 species and varieties found in the state
Calopogon oklahomensis, Oklahoma grass-pink
Calopogon tuberosus, common grass-pink
Corallorhiza odontorhiza var. *odontorhiza*, autumn coralroot
Corallorhiza wisteriana, Wister's coralroot
Epipactis gigantea, stream orchid
Gymnadeniopsis clavellata, little club-spur orchis
Gymnadeniopsis nivea, snowy orchis

Habenaria repens, water-spider orchis
Hexalectris spicata, crested coralroot
Isotria verticillata, large whorled pogonia
Listera australis, southern twayblade
Malaxis unifolia, green adder's-mouth
Platanthera ciliaris, orange fringed orchis
Platanthera cristata, orange crested orchis
Platanthera flava var. *flava*, southern tubercled orchis
[*Platanthera flava* var. *herbiola*, northern tubercled orchis]
Pogonia ophioglossoides, rose pogonia
Spiranthes cernua, nodding ladies'-tresses
Spiranthes lacera var. *gracilis*, southern slender ladies'-tresses
Spiranthes magnicamporum, Great Plains ladies'-tresses
Spiranthes odorata, fragrant ladies'-tresses
Spiranthes ovalis var. *ovalis*, southern oval ladies'-tresses
Spiranthes ovalis var. *erostellata*, northern oval ladies'-tresses
Spiranthes sylvatica, woodland ladies'-tresses
Spiranthes tuberosa, little ladies'-tresses
Spiranthes vernalis, grass-leaved ladies'-tresses
Tipularia discolor, crane-fly orchis
Triphora trianthophora, three birds orchid

Wisconsin

34 species and varieties found in the prairies and plains regions, of 53 species and varieties found in the state
Aplectrum hyemale, putty-root
Calopogon oklahomensis, Oklahoma grass-pink
Calopogon tuberosus, common grass-pink
Corallorhiza maculata var. *maculata*, spotted coralroot
Corallorhiza maculata var. *occidentalis*, western spotted coralroot
Corallorhiza odontorhiza var. *odontorhiza*, autumn coralroot
Corallorhiza odontorhiza var. *pringlei*, Pringle's autumn coralroot
Corallorhiza trifida, early coralroot
Cypripedium acaule, pink lady's-slipper
Cypripedium candidum, small white lady's-slipper
Cypripedium parviflorum var. *makasin*, northern small yellow lady's-slipper
Cypripedium parviflorum var. *pubescens*, large yellow lady's-slipper
Cypripedium reginae, showy lady's-slipper
Galearis spectabilis, showy orchis
Goodyera pubescens, downy rattlesnake orchis
Gymnadeniopsis clavellata, little club-spur orchis

Liparis liliifolia, lily-leaved twayblade
Liparis loeselii, Loesel's twayblade
Malaxis unifolia, green adder's-mouth
Platanthera aquilonis, northern green bog orchis
Platanthera dilatata var. *dilatata*, tall white northern bog orchis
Platanthera hookeri, Hooker's orchis
Platanthera huronensis, green bog orchis
Platanthera lacera, green fringed orchis
Platanthera leucophaea, eastern prairie fringed orchis
Platanthera psycodes, small purple fringed orchis
Pogonia ophioglossoides, rose pogonia
Spiranthes casei, Case's ladies'-tresses
Spiranthes cernua, nodding ladies'-tresses
Spiranthes lacera var. *lacera*, northern slender ladies'-tresses
Spiranthes lacera var. *gracilis*, southern slender ladies'-tresses
Spiranthes magnicamporum, Great Plains ladies'-tresses
Spiranthes romanzoffiana, hooded ladies'-tresses
Triphora trianthophora, three birds orchid

Wyoming

10 species and varieties found in the prairies and plains regions, of 29 species and varieties found in the state
Corallorhiza striata var. *vreelandii*, Vreeland's striped coralroot
Cypripedium parviflorum var. *pubescens*, large yellow lady's-slipper
Goodyera oblongifolia, giant rattlesnake orchis
Goodyera repens, dwarf rattlesnake orchis
Liparis loeselii, Loesel's twayblade
Platanthera aquilonis, northern green bog orchis
Platanthera dilatata var. *dilatata*, tall white northern bog orchis
Platanthera orbiculata, pad-leaved orchis
Platanthera praeclara, western prairie fringed orchis
Spiranthes diluvialis, Ute ladies'-tresses

Some Regional Orchid Statistics

Of the 73 species and varieties of orchids found within the prairies and Great Plains region of North America...

9 species and varieties are typically found in or restricted to the prairies and plains, i.e., grasslands and fens or open wetlands
Calopogon oklahomensis, Oklahoma grass-pink
Calopogon tuberosus, common grass-pink
Cypripedium candidum, small white lady's-slipper
Platanthera lacera, green fringed orchis
Platanthera leucophaea, eastern prairie fringed orchis
Platanthera praeclara, western prairie fringed orchis
Pogonia ophioglossoides, rose pogonia
Spiranthes magnicamporum, Great Plains ladies'-tresses
Spiranthes tuberosa, little ladies'-tresses

24 species and varieties, although found within the grasslands of the prairies and plains, are equally and often more typically at home in other habitats—open woodlands, bogs, swamps, riparian areas, disturbed roadsides
Cypripedium parviflorum var. *makasin*, northern small yellow lady's-slipper
Cypripedium parviflorum var. *pubescens*, large yellow lady's-slipper
Gymnadeniopsis clavellata, little club-spur orchis
Gymnadeniopsis nivea, snowy orchis
Liparis loeselii, Loesel's twayblade
Listera australis, southern twayblade
Malaxis unifolia, green adder's-mouth
Platanthera aquilonis, northern green bog orchis
Platanthera blephariglottis, northern white fringed orchis
Platanthera ciliaris, orange fringed orchis
Platanthera cristata, orange crested orchis
Platanthera flava var. *flava*, southern tubercled orchis
Platanthera flava var. *herbiola*, northern tubercled orchis
Platanthera huronensis, green bog orchis
Platanthera psycodes, small purple fringed orchis
Spiranthes cernua, nodding ladies'-tresses
Spiranthes diluvialis, Ute ladies'-tresses
Spiranthes lacera var. *lacera*, northern slender ladies'-tresses
Spiranthes lacera var. *gracilis*, southern slender ladies'-tresses

Spiranthes laciniata, lace-lipped ladies'-tresses
Spiranthes lucida, shining ladies'-tresses
Spiranthes odorata, fragrant ladies'-tresses
Spiranthes romanzoffiana, hooded ladies'-tresses
Spiranthes vernalis, grass-leaved ladies'-tresses

10 species and varieties are found within the prairie and Great Plains region in only one state or province (additional species may be found in other habitats in the states or provinces treated within this book)
Calypso bulbosa var. *americana*, eastern fairy slipper—South Dakota
Cypripedium acaule, pink lady's-slipper—Manitoba
Goodyera oblongifolia, giant rattlesnake orchis—South Dakota
Goodyera repens, lesser rattlesnake orchis—South Dakota
Listera convallarioides, broad-lipped twayblade—South Dakota
Malaxis brachypoda, white adder's-mouth—Illinois
Piperia unalascensis, Alaskan piperia—South Dakota
Platanthera dilatata var. *albiflora*, bog candles—South Dakota
Platanthera stricta, slender bog orchis—South Dakota
Spiranthes casei, Case's ladies'-tresses—Wisconsin

4 species and varieties are found within the prairie and Great Plains region in only two states or provinces (additional species may be found in other habitats in the states or provinces treated within this book)
Corallorhiza odontorhiza var. *pringlei*, Pringle's autumn coralroot—Wisconsin, Iowa
Cypripedium kentuckiense, ivory-lipped lady's-slipper—Arkansas, Oklahoma
Isotria medeoloides, small whorled pogonia—Missouri, Illinois
Spiranthes diluvialis, Ute ladies'-tresses—Nebraska, Colorado

2 species and varieties are found in three provinces or states in the prairie and Great Plains region
Corallorhiza trifida, early coralroot—Saskatchewan, Manitoba, North Dakota
Spiranthes ovalis var. *ovalis*, southern oval ladies'-tresses—Arkansas, Oklahoma, Louisiana

22 species and varieties are typically northern or northwestern in their distribution, many reaching the southern limit of their range in the prairie and Great Plains region
Calypso bulbosa var. *americana*, eastern fairy slipper
Coeloglossum viride var. *virescens*, long-bracted green orchis
Corallorhiza striata var. *striata*, striped coralroot
Corallorhiza striata var. *vreelandii*, Vreeland's striped coralroot
Corallorhiza trifida, early coralroot
Cypripedium parviflorum var. *makasin*, northern small yellow lady's-slipper

Cypripedium reginae, showy lady's-slipper
Goodyera oblongifolia, giant rattlesnake orchis
Goodyera repens, lesser rattlesnake orchis
Liparis loeselii, Loesel's twayblade
Listera convallarioides, broad-lipped twayblade
Malaxis brachypoda, white adder's-mouth
Piperia unalascensis, Alaskan piperia
Platanthera aquilonis, northern green bog orchis
Platanthera dilatata var. *dilatata*, tall white northern bog orchis
Platanthera dilatata var. *albiflora*, bog candles
Platanthera hookeri, Hooker's orchis
Platanthera huronensis, green bog orchis
Platanthera orbiculata, pad-leaved orchis
Platanthera stricta, slender bog orchis
Spiranthes lacera var. *lacera*, northern slender ladies'-tresses
Spiranthes romanzoffiana, hooded ladies'-tresses

10 species and varieties are typically southern or southeastern in their distribution and reach the northern limit of their distribution in the prairies and Great Plains region

Cypripedium kentuckiense, ivory-lipped lady's-slipper
Cypripedium parviflorum var. *parviflorum*, southern small yellow lady's-slipper
Gymnadeniopsis nivea, snowy orchis
Habenaria repens, water-spider orchis
Hexalectris spicata, crested coralroot
Platanthera cristata, orange crested orchis
Platanthera flava var. *flava*, southern tubercled orchis
Spiranthes odorata, fragrant ladies'-tresses
Spiranthes ovalis var. *ovalis*, southern oval ladies'-tresses
Spiranthes sylvatica, woodland ladies'-tresses

24 species and varieties are typically eastern in their distribution and reach the western limit of their range in the prairies and Great Plains region

Aplectrum hyemale, putty-root
Calopogon tuberosus, common grass-pink
Cypripedium acaule, pink lady's-slipper
Cypripedium reginae, showy lady's-slipper
Goodyera pubescens, downy rattlesnake orchis
Isotria medeoloides, small whorled pogonia
Isotria verticillata, large whorled pogonia
Liparis liliifolia, lily-leaved twayblade
Listera australis, southern twayblade
Malaxis unifolia, green adder's-mouth
Platanthera blephariglottis, northern white fringed orchis

    Platanthera flava var. *herbiola*, northern tubercled orchis
    Platanthera lacera, green fringed orchis
    Platanthera psycodes, small purple fringed orchis
    Pogonia ophioglossoides, rose pogonia
    Spiranthes cernua, nodding ladies'-tresses
    Spiranthes lacera var. *gracilis*, southern slender ladies'-tresses
    Spiranthes lacera var. *lacera*, northern slender ladies'-tresses
    Spiranthes lucida, shining ladies'-tresses
    Spiranthes ovalis var. *erostellata*, northern oval ladies'-tresses
    Spiranthes tuberosa, little ladies'-tresses
    Spiranthes vernalis, grass-leaved ladies'-tresses
    Tipularia discolor, crane-fly orchis
    Triphora trianthophora, three birds orchid

5 species and varieties are typically western in their distribution and reach the eastern limit of their primary range in the prairies and plains
    Corallorhiza striata var. *vreelandii*, Vreeland's striped coralroot
    Epipactis gigantea, stream orchid
    Piperia unalascensis, Alaskan piperia
    Platanthera stricta, slender bog orchid
    Spiranthes diluvialis, Ute ladies'-tresses

4 species are federally listed in the Endangered Species Act
    Isotria medeoloides, small whorled pogonia: Threatened
    Platanthera leucophaea, eastern prairie fringed orchis: Threatened
    Platanthera praeclara, western prairie fringed orchis: Threatened
    Spiranthes diluvialis, Ute ladies'-tresses: Threatened

2 species and varieties, although not listed by the Endangered Species Act, may be so rare globally as to be considered threatened
    Calopogon oklahomensis, Oklahoma grass-pink
    Spiranthes ovalis var. *ovalis*, southern oval ladies'-tresses

11 species and varieties have recently been described or revalidated
    Calopogon oklahomensis, Oklahoma grass-pink
    Corallorhiza odontorhiza var. *pringlei*, Pringle's autumn coralroot
    Corallorhiza maculata var. *occidentalis*, western spotted coralroot
    Cypripedium kentuckiense, ivory-lipped lady's-slipper
    Platanthera aquilonis, northern green bog orchis
    Platanthera praeclara, western prairie fringed orchis
    Spiranthes casei, Case's ladies'-tresses
    Spiranthes diluvialis, Ute ladies'-tresses
    Spiranthes magnicamporum, Great Plains ladies'-tresses
    Spiranthes ovalis var. *erostellata*, northern oval ladies'-tresses
    Spiranthes sylvatica, woodland ladies'-tresses

Rare, Threatened, and Endangered Species

In the United States each state has a somewhat different system for listing and ranking rare plants, whereas the Natural Heritage Programs and NatureServe have a uniform system for ranking species. The list of the orchids for each state and province may also be influenced by the presence, or lack thereof, of persons with a particular interest in the native orchids. The following rankings and their definitions should enable the reader to understand the current status of the listed species. This information is taken from material provided by credited sources and printed exactly as they list the species, with current synonyms in parentheses.

- S1   Extremely rare throughout its range in the state or province (typically 5 or fewer occurrences or very few remaining individuals). May be especially vulnerable to extirpation.
- S2   Rare throughout its range in the state or province (6–20 occurrences or few remaining individuals). May be vulnerable to extirpation because of rarity or other factors.
- S3   Uncommon throughout its range in the state or province, or found only in a restricted range, even if abundant at some locations. (21–100 occurrences).
- S#S#   Numeric range rank: A range between two consecutive numeric ranks. Denotes uncertainty about the exact rarity of the species; e.g., S1S2.
- SH   Historical: Species occurred historically throughout its range in the state or province (with expectation that it may be rediscovered), perhaps having not been verified in the past 20–70 years (depending on the species), and suspected to be still extant.

Provincial and state rankings indicate the relative rarity of the species within the province or state only. National and global rankings indicate the rarity within a larger area. Most species within the scope of this field guide are ranked as G4 or G5, indicating that globally they are considered secure. Only 4 species are ranked otherwise.

Sources:
NatureServe
http://www.natureserve.org/explorer/servlet/NatureServe?loadTemplate-tabular_report.wmt&paging-home&save-all&sourceTemplate-reviewMiddle.wmt
Committee on the Status of Endangered Wildlife in Canada
http://www.cosewic.gc.ca

SU Unconfirmed record; usually from the out-of-area literature. Unrankable: Possibly in peril throughout its range in the state or province, but status uncertain; need more information.

SX Extinct/Extirpated: Species is believed to be extirpated within the state or province.

S? Unranked: Species is not yet ranked.

GH Of historical occurrence throughout its range; may be rediscovered.

GX Believed to be extinct throughout its range.

Additional ranks are used for those species whose populations are ample and secure, and not noted here.

Note: Because of recent taxonomic work involving *Corallorhiza maculata* var. *maculata*/*C. maculata* var. *occidentalis* and *Platanthera aquilonis*/*P. huronensis* precise distributional information is still not complete. None of these species or varieties is exceptionally rare, but they need to be sorted out to determine relative rarity within each province.

Manitoba

Calopogon tuberosus var. *tuberosus*, Grass-pink	S2
Corallorhiza striata var. *striata*, Striped coralroot	S3?
Cypripedium candidum, Small white lady's-slipper	S1
Cypripedium reginae, Showy lady's-slipper	S3?
Liparis loeselii, Loesel's false-twayblade	S3?
Malaxis unifolia, Green adder's-mouth	S2?
Platanthera hookeri, Hooker's rein-orchid	S2
Platanthera lacera, Ragged fringed-orchid	S2
Platanthera orbiculata, Pad-leaved rein orchid	S3
Platanthera praeclara, Western prairie fringed-orchid	S1
Platanthera psycodes, Small purple fringed-orchid	S1
Pogonia ophioglossoides, Rose pogonia	S1
Spiranthes magnicamporum, Great plains ladies'-tresses	S1?

Source:
Native Orchid Conservation, Inc.
http://www.nativeorchid.com/mborchids.htm

Saskatchewan

Cypripedium candidum, small white lady's-slipper	SX

Source:
Saskatchewan Environment
http://www.se.gov.sk.ca/ecosystem/speciesatrisk/

Arkansas

Calopogon oklahomensis, Oklahoma grass-pink	S2
Calopogon tuberosus, common grass-pink	S2
Cypripedium kentuckiense, ivory-lipped lady's-slipper	S3
Cypripedium reginae, showy lady's-slipper	G4 S1 SE
Habenaria repens, water spider orchid	S2
Hexalectris spicata, crested coralroot	S2
Liparis loeselii, Loesel's twayblade	S1
Platanthera cristata, orange crested orchis	S1S2
Platanthera flava, southern tubercled orchis	S1S2
Platanthera nivea, snowy orchis (*Gymnadeniopsis nivea*)	SH
Platanthera peramoena, purple fringeless orchis	S2
Pogonia ophioglossoides, rose pogonia	S2
Spiranthes lucida, shining ladies'-tresses	G5 S2
Spiranthes magnicamporum, Great Plains ladies'-tresses	S1?
Spiranthes odorata, fragrant ladies'-tresses	S1
Spiranthes praecox, giant ladies'-tresses (*Spiranthes sylvatica* woodland ladies'-tresses)	S1S2

Source:
Cindy Osborne, Data Manager
Arkansas Natural Heritage Commission
1500 Tower Building, 323 Center Street
Little Rock, AR 72201
Phone: 501-324-9762, Fax: 501-324-9618
e-mail: cindy@dah.state.ar.us
http://www.heritage.state.ar.us:80/nhc/heritage.html

Colorado

Cypripedium calceolus subsp. *parviflorum* (*C. parviflorum* var. *pubescens*), large yellow lady's-slipper	G5/S2
Epipactis gigantea, giant helleborine	G4/S2
Goodyera repens, Dwarf rattlesnake-plantain	G5/S2
Spiranthes diluvialis, Ute ladies'-tresses	G2/S2 FT

Source:
Colorado Natural Heritage Program
254 General Services Building
Colorado State University
Fort Collins, CO 80523
970-491-2992
http://www.cnhp.colostate.edu/rareplants/masterlist.hml

Illinois

Endangered

Calopogon oklahomensis, Oklahoma grass-pink
Calopogon tuberosus, common grass-pink
Cypripedium acaule, pink lady's-slipper, moccasin flower
Cypripedium parviflorum var. *makasin*, northern small yellow lady's-slipper
Cypripedium reginae, showy lady's-slipper
Hexalectris spicata, crested coralroot
Isotria verticillata, large whorled pogonia
Platanthera ciliaris, orange fringed orchis
Platanthera clavellata (*Gymnadeniopsis clavellata*), little club-spur orchis
Platanthera flava var. *flava*, southern tubercled orchis
Platanthera leucophaea, eastern prairie fringed orchid
Platanthera psycodes, small purple fringed orchis
Spiranthes lucida, shining ladies'-tresses
Spiranthes vernalis, grass-leaved ladies'-tresses

Threatened

Corallorhiza maculata, spotted coralroot
Cypripedium candidum, small white lady's-slipper
Platanthera flava var. *herbiola*, northern tubercled orchis

Source:
Illinois Endangered Species Protection Board
One Natural Resources Way
Springfield, Illinois 62702–1271 http://dnr.state.il.us/espb/datelist.htm

Iowa

Endangered

Platanthera flava, tubercled orchis
Platanthera leucophaea, eastern prairie fringed orchid
Spiranthes lucida, shining ladies'-tresses

Threatened

Corallorhiza maculata, spotted coralroot
Cypripedium reginae, showy lady's-slipper
Platanthera hookeri, Hooker's orchis
Platanthera hyperborea, (*P. aquilonis*), northern leafy green bog orchis
Platanthera praeclara, western prairie fringed orchid
Platanthera psycodes, purple fringed orchid

Spiranthes lacera, slender ladies'-tresses
Spiranthes ovalis, oval ladies-tresses
Spiranthes romanzoffiana, hooded ladies-tresses
Spiranthes vernalis, spring ladies-tresses

SPECIAL CONCERN

Calopogon tuberosus, common grass-pink
Platanthera clavellata (*Gymnadeniopsis*), little club-spur orchis
Platanthera lacera, green fringed orchid
Spiranthes magnicamporum, Great Plains ladies'-tresses

Source:
Iowa Department of Natural Resources
Threatened and Endangered Species
http://www.iowadnr.com/other/threatened.html

Kansas

Calopogon tuberosus tuberous grass-pink	g5 s1
Calopogon oklahomensis, Oklahoma grass-pink	g4? S1
Hexalectris spicata, crested coralroot	g5 s1
Liparis loeselii, Loesel's twayblade	g5 sx
Malaxis unifolia, green adder's-mouth	g5 s1
Platanthera lacera, green-fringed orchis	g5 s2
Platanthera praeclara, western prairie fringed orchid	g2 s1
Spiranthes lucida, shining ladies'-tresses	g5 sh
Spiranthes ovalis, lesser ladies'-tresses	g5? S1
Triphora trianthophora, nodding pogonia	g3g4 s2

Source:
Kansas Biological Survey
2101 Constant Avenue, Higuchi Hall
The University of Kansas
Lawrence, KS 66047-3759
Phone: 785-864-1500, Fax: 785-864-1534

Louisiana

Calopogon barbatus, bearded grass-pink (includes plants now known as *Calopogon oklahomensis*)	s1
Corallorhiza odontorhiza, autumn coralroot	s1
Cypripedium kentuckiense, ivory-lipped lady's-slipper	s1

Isotria verticillata, large whorled pogonia   s2s3
Platanthera lacera, green fringed orchis, ragged orchis   s1
Spiranthes magnicamporum, Great Plains ladies'-tresses   s1
Triphora trianthophora, three birds orchid   s1

Source: David Brunet
Brunet_DP@wlf.state.la.us

Minnesota

Endangered

Platanthera flava var. *herbiola*, tubercled rein-orchid
Platanthera praeclara, western prairie fringed orchid

Threatened

Cypripedium candidum, small white lady's-slipper
Platanthera clavellata (*Gymnadeniopsis clavellata*), club-spur orchid

Source:
Natural Heritage and Nongame Research Program
Minnesota Department of Natural Resources
Section of Ecological Services
500 Lafayette Rd., Box 25
St. Paul, MN 55155
Phone: 1-800-766-6000, Fax: 651-296-1811
http://files.dnr.state.mn.us/natural_resources/ets/endlist.pdf

Missouri

Cypripedium candidum, small white lady's-slipper	S1
Cypripedium reginae, showy lady's-slipper	S2S3
Isotria medeoloides, small whorled pogonia	SH
Isotria verticillata, large whorled pogonia	S1S2
Liparis loeselii, Loesel's twayblade	S2
Platanthera ciliaris, orange fringed orchis	S1
Platanthera clavellata (*Gymnadeniopsis clavellata*), little club-spur orchis	S2
Platanthera flava var. *flava*, southern tubercled orchis	S2
Platanthera flava var. *herbiola*, northern tubercled orchis	S2
Platanthera leucophaea, eastern prairie fringed orchis	SH
Platanthera praeclara, western prairie fringed orchis	S1
Platanthera psycodes, small purple fringed orchis	SX
Pogonia ophioglossoides, rose pogonia	S1
Tipularia discolor, crane-fly orchis	S1

Source:
Tim E. Smith, Botanist
Missouri Department of Conservation
P.O. Box 180
Jefferson City, MO 65102-0180
Phone: 573-522-4115 ext. 3200, Fax: 573-526-5582
Tim.Smith@mdc.mo.gov

Montana

Cypripedium parviflorum var. *pubescens*, large yellow lady's-slipper	S2/S3
Epipactis gigantea, stream orchid	S2
Goodyera repens, lesser rattlesnake orchis	S2/S3
Liparis loeselii, Loesel's twayblade or fen orchis	S1

Source:
Montana Natural Heritage Program
http://nhp.nris.state.mt.us/plants/index.html

Nebraska

Platanthera praeclara, western prairie fringed orchid	Threatened
Spiranthes diluvialis, Ute ladies'-tresses	Threatened

Source:
Endangered Specie.com
http://www.endangeredspecie.com/states/ne.htm

New Mexico

Cypripedium parviflorum, yellow lady's-slipper	G5 S? No 0
Corallorhiza striata, striped coral-root	D G5 S4? No 1
Corallorhiza striata var. *striata*, striped coral-root	G5T4T5 S3 No 0
Cypripedium parviflorum var. *pubescens*, yellow lady's-slipper	E G5T4T5 S2?
Epipactis gigantea, giant helleborine orchid	G3 S2?
Habenaria dilatata (*Platanthera dilatata*) white bog orchid	G5 S2
Habenaria (*Piperia*) *unalascensis*, Alaska bog-orchid	G5 S1
Spiranthes magnicamporum, Great Plains ladies'-tresses	E G4 S3?
Spiranthes romanzoffiana, hooded ladies'-tresses	G5 S2?

Source:
New Mexico Natural Heritage Program
Museum of Southwestern Biology
Department of Biology

University of New Mexico
Albuquerque, NM 87131
http://nmnhp.unm.edu

North Dakota

North Dakota does not have a list of rare, threatened, or endangered species

Oklahoma

Aplectrum hyemale, puttyroot	g5 s1
Calopogon oklahomensis, Oklahoma grass-pink	g4? sr
Calopogon tuberosus, tuberous grass-pink	g5 s2s3
Corallorhiza odontorhiza, autumn coral-root	g5 s1
Cypripedium calceolus (*C. parviflorum* var. *parviflorum*), European yellow lady's-slipper	g4? s2
Cypripedium kentuckiense, southern lady's-slipper	g3 s1
Epipactis gigantea, giant helleborine	g3 s1s2
Habenaria repens, water-spider orchid	g5 s1
Hexalectris spicata, crested coralroot	g5 s1s2
Isotria verticillata, large whorled pogonia	g5 s1
Liparis liliifolia, large twayblade	g5 s1
Listera australis, southern twayblade	g4 s1
Malaxis unifolia, green adder's-mouth	g5 s1
Platanthera ciliaris, yellow-fringed orchid	g5 s1
Platanthera clavellata (*Gymnadeniopsis clavellata*), small green woodland orchid	g5 s1s2
Platanthera flava, southern rein-orchid	g4 s1
Platanthera lacera, green-fringe orchis	g5 s1s2
Platanthera leucophaea, eastern prairie white-fringed orchid	g2 sh
Platanthera praeclara, western prairie fringed orchid	g2 s1
Pogonia ophioglossoides, rose pogonia	g5 s1
Spiranthes odorata, sweetscent ladies'-tresses	g5 s1
Spiranthes praecox, grassleaf ladies'-tresses (*Spiranthes sylvatica*)	g5 s1
Tipularia discolor, cranefly orchid	g4g5 s1
Triphora trianthophora, nodding pogonia	g3g4 s2s3

Source:
Oklahoma Biological Survey
http://www.biosurvey.ou.edu/fedspp.html

South Dakota

Calypso bulbosa, fairy slipper orchid	G5 S3
Corallorhiza odontorhiza, autumn coral-root	G5 SU
Corallorhiza trifida, pale coral-root	G5 S2
Cypripedium calceolus, yellow lady's-slipper	G5 S3?
Cypripedium candidum, small white lady's-slipper	G4 S1
Epipactis gigantea, stream orchid	G3 S1
Liparis loeselii, Loesel's twayblade	G5 S1
Listera convallarioides, broad-lipped twayblade	G5 S1
Platanthera dilatata, northern white orchid	G5 SU
Platanthera orbiculata, round-leaved orchid	G5? S2
Platanthera praeclara, western prairie fringed orchid	LT G2 SH
Spiranthes cernua, nodding ladies' tresses	G5 S2
Spiranthes magnicamporum, Great Plains ladies' tresses	G4 SU
Spiranthes vernalis, twisted ladies' tresses	G5 S2

Source:
The South Dakota Natural Heritage Program
South Dakota Department of Game, Fish and Parks
http://www.sdgfp.info/Wildlife/Diversity/rareplant2002.htm

Texas

No orchids found within the plains and prairies region of Texas are listed as state endangered, threatened, or rare.

Source:
Texas Parks and Wildlife Department
4200 Smith School Road
Austin, TX 78744
http://www.tpwd.state.tx.us/nature/endang/regulations/texas/

Wisconsin

Cypripedium candidum, small white lady's-slipper	G4 S3 THR
Platanthera flava var. *herbiola*, pale green orchid	G4T4Q S2 THR
Platanthera leucophaea, prairie white-fringed orchid	G2 S3 LT END

Source:
Wisconsin DNR Central Office
101 S. Webster Street
P.O. Box 7921
Madison, WI 53707-7921

608-266-2621
http://www.dnr.state.wi.us/org/land/er/working_list/taxalists/TandE.htm

Wyoming

Cypripedium parviflorum var. *pubescens*, large yellow lady-slipper G5T?/ S1S2
Platanthera orbiculata, round-leaved orchid G5?/ S1
Spiranthes diluvialis, Ute ladies' tresses G3/S1

Source:
Keinath, D., B. Heidel, and G. P. Beauvais. 2003. Wyoming Plant and Animal Species of Concern.
Wyoming Natural Diversity Database
University of Wyoming
Laramie, WY
http://uwadmnweb.uwyo.edu/WYNDD/Plants/2003_Plant_SOC.pdf

Recent Literature References for New Taxa, Combinations, and Additions

Note: references for recently described forms are found within the species treatments. Additional information may be found in volume 26 of *Flora of North America* (FNA, 1993).

Calopogon oklahomensis D.H. Goldman
Oklahoma grass-pink
Goldman, D. H. 1995. *Lindleyana* 10(1): 37–42.

Corallorhiza maculata (Rafinesque) Rafinesque var. *occidentalis* (Lindley) Ames
western spotted coralroot
Freudenstein, J. V. 1986. *Contributions from the University of Michigan Herbarium* 16: 145–153.
———. 1997. *Harvard Papers in Botany* 10: 5–51.
Corallorhiza odontorhiza (Willdenow) Nuttall var. *pringlei* (Greenman) Freudenstein
Pringle's autumn coralroot
Freudenstein, J. V. 1993. Dissertation, Cornell University.
———. 1997. *Harvard Papers in Botany* 10: 5–51.

Cypripedium kentuckiense C.F. Reed
ivory-lipped lady's-slipper
Atwood, J. T., Jr. 1984. *AOS Bulletin* 53(8): 835–41.
Brown, P.M. 1995. *NANOJ* 1(3): 255.
———. 1998. *NANOJ* 4(1): 45.
Reed, C. 1981. *Phytologia* 48(5): 426–28.
Weldy, T. W., H. T. Mlodozeniec, L. E. Wallace, and M. A. Case. 1996. *Sida* 17(2): 423–35.
Cypripedium parviflorum Salisbury var. *parviflorum*
southern small yellow lady's-slipper
Sheviak, C. J. 1994. *AOS Bulletin* 63(6): 664–69.
———. 1995. *AOS Bulletin* 64(6): 606–12.
———. 1996. *NANOJ* 2(4): 319–43.
Cypripedium parviflorum Salisbury var. *makasin* (Farwell) Sheviak
northern small yellow lady's-slipper
Sheviak, C. J. 1993. *AOS Bulletin* 62(4): 403.

———. 1994. *AOS Bulletin* 63(6): 664–69.
———. 1995. *AOS Bulletin* 64(6): 606–12.
———. 1996. *NANOJ* 2(4): 319–43.
Cypripedium parviflorum Salisbury var. *pubescens*
large yellow lady's-slipper
Sheviak, C. J. 1994. *AOS Bulletin* 63(6): 664–69.
———. 1995. *AOS Bulletin* 64(6): 606–12.
———. 1996. *NANOJ* 2(4): 319–43.

Gymnadeniopsis clavellata (Michaux) Rydberg
Gymnadeniopsis integra (Nuttall) Rydberg
Gymnadeniopsis nivea (Nuttall) Rydberg
Brown, P. M. 2002. *North American Native Orchid Journal* 8: 32–40.

Platanthera aquilonis Sheviak
northern green bog orchis
Sheviak, C. J. 1999. *Lindleyana* 14(4): 193–203
Wallace, L. E. 2004. *Canadian Journal of Botany* 82: 244–52.
Platanthera huronensis (Nuttall) Lindley
green bog orchis
Wallace, L. E. 2003. *International Journal of Plant Science*. 164(6): 907–16.
———. 2004. *Canadian Journal of Botany* 82: 244–52.
Platanthera leucophaea (Nuttall) Lindley
eastern prairie fringed orchis
Sheviak, C. J., and M. Bowles. 1986. *Rhodora* 88: 267–90.
Platanthera peramoena (A. Gray) A. Gray
purple fringeless orchis
Spooner, D. M., and J. S. Shelly. 1983. *Rhodora* 85: 55–64.
Platanthera praeclara Sheviak & Bowles
western prairie fringed orchis
Sheviak, C. J., and M. Bowles. 1986. *Rhodora* 88: 267–90.
Platanthera psycodes (Linnaeus) Lindley
small purple fringed orchis
Stoutamire, W. P. 1974. *Brittonia* 26: 42–58.

Spiranthes casei Catling & Cruise var. *casei*
Case's ladies'-tresses
Catling, P. M., and J. E. Cruise. 1974. *Rhodora* 76: 256–536.
Spiranthes cernua (Linnaeus) L.C. Richard
nodding ladies'-tresses
Sheviak, C. J. 1991. *Lindleyana* 6(4): 228–34.
Spiranthes diluvialis Sheviak
Ute ladies'-tresses
———. 1991. *Lindleyana* 6(4): 228–34.

Spiranthes magnicamporum Sheviak
Great Plains ladies'-tresses
———. 1973. *Botanical Museum Leaflet of Harvard University* 23: 285–97.
Spiranthes ovalis Lindley var. *erostellata* Catling
northern oval ladies'-tresses
Catling, P. M. 1983. *Brittonia* 35: 120–25.
Spiranthes sylvatica P.M. Brown
woodland ladies'-tresses
Brown, P. M. 2001. *North American Native Orchid Journal* 7(3): 193–205.

Synonyms and Misapplied Names

Synonyms and misapplied names are both often confused and confusing in the literature and in the understanding of orchid enthusiasts. A synonym is simply an alternate name for a previously published plant name. From among the synonyms each author must select a name that he or she feels best suits the currently accepted genus and species for a given plant. Although the genus may vary, the species epithet may often be the same. In such large groups as the spiranthoid orchids (*Spiranthes* and its allied genera) many synonyms may exist for the same species—all within different genera.

The rules of priority, as set forth in the *International Code of Botanical Nomenclature* (2000), dictate that the earliest validly published species name within the chosen genus must be used. A good example would be that of the common grass-pink, *Calopogon tuberosus* (Linnaeus) Britton, Sterns, & Poggenberg. This species was originally published by Linnaeus in 1753 as *Limodorum tuberosum* and the new combination by Britton et al. was made in 1888. In the interim *Calopogon pulchellus* (Salisbury) R. Brown (1813) was based on *Limodorum pulchellum* Salisbury (1796). Although the name *Calopogon pulchellus* was in widespread usage for many years, priority indicates that the earliest validly published name must be used. Because Linnaeus's specific epithet *tuberosum* predates Salisbury's *pulchellum*, *tuberosum* must be used in the new combination with *Calopogon*. The endings are dictated by the case/gender of the generic name. Usually the year of publication is sufficient for determining the valid name, though on rare occasions it has come down to mere months between publications.

Orchis spectabilis would be a synonym for *Galearis spectabilis*, *Galearis* being the currently accepted genus for the North American species. *Habenaria* is another group that has undergone a great deal of scrutiny in the past twenty-five years. Several groups of species formerly included within *Habenaria* are now treated as distinct genera. This is not always so much a case of correct or incorrect names, but that of the preference of the author for one genus over another. The recent trend for molecular, i.e., DNA, analysis of species has resulted in the placement of several species or groups of species within other genera. Although this is tempting to accept as the end-all and be-all of taxonomy, it is still only one tool to help in formulating an opinion.

A misapplied name is an incorrect name for a given plant that may have resulted from a reassessment of the genus or species, resulting in two or more species being described from within the original species, or it may simply be a wrong name assigned to the plants. This is especially common in geographic areas at the edge of a group's range. The most frequently encountered example would be *Cypripedium calceolus*. For many years our North American yellow lady's-slippers have been treated as a geographic variant of the Eurasian species. Many years of work by Sheviak have demonstrated that this is not the case and that the North American plants are a distinct species—*Cypripedium parviflorum*. Therefore, the name *C. calceolus* is a misapplied name for the North American plants. The term *auct.* (*auctorum*, i.e., of authors) is used to indicate a misapplied name, and occasionally an author will incorrectly append the phrase "in part" after a name listed under synonymy. Misapplied names are not synonyms and refer only to the specific geographic area being treated—in this case the prairies and Great Plains of North America.

An issue can arise as to whether a name is a synonym or misapplied, and that depends on a broad or narrow view of the taxonomy—the lumpers vs. the splitters. Such a situation would best be described thus: if *Platanthera huronensis* is considered synonymous with *Platanthera hyperborea*, *Platanthera huronensis* becomes a synonym of *Platanthera hyperborea*; but if *Platanthera huronensis* is considered a good species on its own, *Platanthera hyperborea* becomes a misapplied name for *Platanthera huronensis*. At times this appears to be an endless argument and each author must make a decision as to synonymy and misapplied names. In the prairies and plains regions of North America there are very few misapplied names. Cross-references to all taxa with synonyms and misapplied names can be found at the end of this chapter.

Synonyms and misapplied names are given for taxa found in the following as well as occasional references to specific journal articles.

Brown, P. M., and S.N. Folsom. 2003. *The Wild Orchids of North America, North of Mexico.*
———. 2004. *Wild Orchids of the Southeastern United States, North of Peninsular Florida.*
Case, F. W. 1987. *Orchids of the Western Great Lakes Region.*
Correll, D. S. 1950. *Native Orchids of North America.*
Flora of North America. 2002. Vol. 26, *Orchidaceae.*
Homoya, M. 1993. *Orchids of Indiana.*
Luer, C. A. 1975. *The Native Orchids of the United States and Canada excluding Florida.*
Scoggin, H. J. 1978. *The Flora of Canada*, Part 2.
Slaughter, C. 1995. *Orchids of Arkansas.*
Smith, W. R. 1993. *Orchids of Minnesota.*
Williams, J. G., A. E. Williams, and N. Arlott. 1983. *A Field Guide to the Orchids of North America.*

Segregate Genera

Five genera within the range of this book have been historically treated within the genus *Habenaria*, and are currently treated differently by various authors. These are:
Coeloglossum
Gymnadeniopsis
Piperia

Platanthera
Habenaria

Synonyms

Most current taxonomic treatments recognize the numerous segregate genera of the spiranthoid orchids (Garay, 1980). References given are for recent works, not necessarily the taxonomic work that first designated the new combination. Volume 26 of the *Flora of North America* (FNA, 2002) treats many of the following taxa as well.

Calopogon tuberosus (Linnaeus) Britton, Sterns, & Poggenberg
SYNONYMS
Calopogon pulchellus (Salisbury) R. Brown
Calopogon pulchellus (Salisbury) R. Brown var. *latifolius* (St. John) Fernald
Limodorum pulchellum Salisbury
Limodorum tuberosum Linnaeus

Calypso bulbosa (Linnaeus) Oakes var. *americana* (R. Brown) Luer
SYNONYM
Cytherea bulbosa (Linnaeus) House *p.p.*

Coeloglossum viride (Linnaeus) Hartman var. *virescens* (Mühlenberg) Luer
SYNONYMS
Coeloglossum bracteatum (Mühlenberg *ex* Willdenow) Parlin
Coeloglossum viride subsp. *bracteatum* (Mühlenberg *ex* Willdenow) Hultén
Dactylorhiza viridis R.M. Bateman, A. Pridgeon, & M.W. Chase
Habenaria bracteata (Mühlenberg *ex* Willdenow) R. Brown *ex* Aiton *f.*
Habenaria viridis Linnaeus var. *bracteata* (Mühlenberg *ex* Willdenow)
    Reichenbach *ex* Gray

Corallorhiza maculata (Rafinesque) Rafinesque var. *maculata*
SYNONYM
Corallorhiza multiflora Nuttall

Corallorhiza maculata (Rafinesque) Rafinesque var. *occidentalis* (Lindley) Ames
SYNONYMS
Corallorhiza maculata (Rafinesque) Rafinesque subsp. *occidentalis* (Lindley)
    Cockerell

Corallorhiza odontorhiza (Willdenow) Nuttall var. *pringlei* (Greenman)
    Freudenstein
SYNONYM
Corallorhiza pringlei Greenman

Corallorhiza striata Lindley var. *vreelandii* (Rydberg) L.O. Williams
SYNONYM
Corallorhiza bigelovii S. Watson

Corallorhiza trifida Chatelain
SYNONYMS
Corallorhiza corallorhiza (Linnaeus) Karsten

Cypripedium acaule Aiton
SYNONYM
Fissipes acaulis (Aiton) Small

Cypripedium kentuckiense C.F. Reed
SYNONYM
C. daultonii V. Soukup *nom. nud.*

Cypripedium parviflorum Salisbury var. *parviflorum*
SYNONYM
Cypripedium calceolus Linnaeus var. *parviflorum* (Salisbury) Fernald *p.p.*

Cypripedium parviflorum Salisbury var. *pubescens* (Willdenow) Knight
SYNONYMS
Cypripedium *pubescens* Willdenow
Cypripedium calceolus Linnaeus var. *planipetalum* (Fernald) Victorin & Rousseau
Cypripedium calceolus Linnaeus var. *pubescens* (Willdenow) Correll
Cypripedium parviflorum var. *planipetalum* Fernald
Cypripedium flavescens de Candolle
Cypripedium veganum Cockerell & Barber
MISAPPLIED
Cypripedium calceolus Linnaeus

Cypripedium reginae Walter
SYNONYM
Cypripedium spectabile Salisbury

Epipactis gigantea Douglas *ex* Hooker
SYNONYMS
Amesia gigantea (Douglas) A. Nelson & Macbride
Helleborine gigantea (Douglas) Druce

Epipactis helleborine (Linnaeus) Cranz*
SYNONYM
Epipactis latifolia (Linnaeus) Allioni*

Goodyera oblongifolia Rafinesque
SYNONYM
Goodyera decipiens (Hooker) F.T. Hubbard

Gymnadeniopsis clavellata (Michaux) Rydberg *ex* Britton
SYNONYMS
Habenaria clavellata (Michaux) Sprengel
Platanthera clavellata (Michaux) Luer

Gymnadeniopsis nivea (Nuttall) Rydberg
SYNONYMS
Habenaria nivea (Nuttall) Sprengel
Platanthera nivea (Nuttall) Luer

Isotria medeoloides (Pursh) Rafinesque
SYNONYMS
Isotria affinis (Austin *ex* A. Gray) Rydberg
Pogonia affinis Austin *ex* A. Gray

Isotria verticillata (Mühlenberg *ex* Willdenow) Rafinesque
SYNONYM
Pogonia verticillata (Mühlenberg *ex* Willdenow) Nuttall

Listera australis Lindley
SYNONYM
Neottia australis (Lindley) Szlachetko

Piperia unalascensis (Sprengel) Rydberg
SYNONYMS
Habenaria unalascensis (Sprengel) S. Watson
Platanthera unalascensis (Sprengel) Kurtz
Spiranthes unalascensis Sprengel
MISAPPLIED
Platanthera foetida Geyer ex Hooker

Platanthera aquilonis Sheviak
MISAPPLIED
Habenaria hyperborea (Linnaeus) R. Brown *ex* Aiton
Platanthera hyperborea (Linnaeus) Lindley

Platanthera blephariglottis (Willdenow) Lindley
SYNONYMS
Blephariglottis blephariglottis (Willdenow) Rydberg
Habenaria blephariglottis (Willdenow) Hooker

Platanthera ciliaris (Linnaeus) Lindley
SYNONYMS
Blephariglottis ciliaris (Linnaeus) Rydberg
Habenaria ciliaris (Linnaeus) R. Brown

Platanthera cristata (Michaux) Lindley
SYNONYMS
Blephariglottis cristata (Michaux) Rafinesque
Habenaria cristata (Michaux) R. Brown

Platanthera dilatata (Pursh) Lindley var. *dilatata*
SYNONYMS
Habenaria dilatata (Pursh) Hooker
Limnorchis dilatata (Pursh) Rydberg *ex* Britton
Piperia dilatata (Pursh) Szlachetko & P. Rutkowski

Platanthera dilatata (Pursh) Lindley var. *albiflora* (Chamisso) Ledebour
SYNONYMS
Habenaria borealis Chamisso var. *albiflora* Chamisso
Habenaria dilatata (Pursh) Hooker var. *albiflora* (Chamisso) Correll
Piperia dilatata (Pursh) Szlachetko & P. Rutkowski var. *albiflora* (Chamisso) Szlachetko & P. Rutkowski

Platanthera flava (Linnaeus) Lindley var. *flava*
SYNONYM
Habenaria flava (Linnaeus) R. Brown

Platanthera flava (Linnaeus) Lindley var. *herbiola* (R. Brown) Luer
SYNONYMS
Habenaria flava var. *herbiola* (R. Brown *ex* Aiton) Ames & Correll
Habenaria flava var. *virescens sensu* Fernald
Habenaria herbiola R. Brown *ex* Aiton
Platanthera huronensis (Nuttall) Lindley

SYNONYMS
Habenaria hyperborea Linnaeus var. *huronensis* (Nuttall) Farwell
Habenaria ×*media* (Rydberg) Niles
Limnorchis media Rydberg
Platanthera hyperborea (Linnaeus) Lindley var. *huronensis* (Nuttall) Luer
Platanthera ×*media* (Rydberg) Luer

Platanthera lacera (Michaux) G. Don
SYNONYM
Habenaria lacera (Michaux) R. Brown

Platanthera leucophaea (Nuttall) Lindley
SYNONYM
Habenaria leucophaea (Nuttall) A. Gray

Platanthera orbiculata (Pursh) Lindley
SYNONYM
Habenaria orbiculata (Pursh) Torrey

Platanthera peramoena (A. Gray) A. Gray
SYNONYMS
Blephariglottis peramoena (Gray) Rydberg
Habenaria peramoena A. Gray

Platanthera praeclara Sheviak & Bowles
SYNONYM
Habenaria leucophaea (Nuttall) A. Gray var. *praeclara* (Sheviak & Bowles) Cronquist

Platanthera psycodes (Linnaeus) Lindley
SYNONYMS
Blephariglottis psycodes (Linnaeus) Rydberg
Habenaria psycodes (Linnaeus) Sprengel

Platanthera stricta Lindley
SYNONYMS
Habenaria borealis Chamisso var. *viridiflora* Chamisso
Habenaria saccata Greene
Limnorchis stricta (Lindley) Rydberg
Platanthera gracilis Lindley
Platanthera hyperborea (Linnaeus) Lindley var. *viridiflora* (Chamisso) Kitamura
Platanthera hyperborea (Linnaeus) Lindley var. *viridiflora* (Chamisso) Luer
Platanthera saccata (Greene) Hultén

Spiranthes cernua (Linnaeus) Richard
SYNONYM
Ibidium cernuum (Linnaeus) House

Spiranthes diluvialis Sheviak
SYNONYM
Spiranthes romanzoffiana Chamisso var. *diluvialis* (Sheviak) S.L. Welsh

Spiranthes lacera Rafinesque var. *gracilis* (Bigelow) Luer
SYNONYMS
Ibidium gracile (Bigelow) House
Ibidium beckii (Lindley) House
Neottia gracilis Bigelow
Spiranthes beckii Lindley
Spiranthes gracilis (Bigelow) Beck

Spiranthes laciniata (Small) Ames
SYNONYM
Ibidium laciniatum (Small) House

Spiranthes odorata (Nuttall) Lindley
SYNONYMS
Ibidium odoratum (Nuttall) House
Spiranthes cernua (Linnaeus) L.C. Richard var. *odorata* (Nuttall) Correll

Spiranthes praecox (Walter) S. Watson
SYNONYM
Ibidium praecox (Walter) House

Spiranthes tuberosa Rafinesque
SYNONYMS
Spiranthes grayi Ames
Spiranthes tuberosa var. *grayi* (Ames) Fernald
Spiranthes beckii Lindley
MISAPPLIED
Ibidium beckii House

Spiranthes vernalis Engelmann & A. Gray
SYNONYM
Ibidium vernale (Engelmann & A. Gray) House

Cross-references for all synonyms

The synonym is given first, followed by the currently acceptable names used in this work.
= synonym
≠ misapplied name
Amesia gigantea (Douglas) A. Nelson & Macbride = *Epipactis gigantea* Douglas *ex* Hooker
Blephariglottis blephariglottis (Willdenow) Rydberg = *Platanthera blephariglottis* (Willdenow) Lindley

Blephariglottis ciliaris (Linnaeus) Rydberg = *Platanthera ciliaris* (Linnaeus) Lindley

Blephariglottis cristata (Michaux) Rafinesque = *Platanthera cristata* (Michaux) Lindley

Blephariglottis peramoena (Gray) Rydberg = *Platanthera peramoena* (A. Gray) A. Gray

Blephariglottis psycodes (Linnaeus) Rydberg = *Platanthera psycodes* (Linnaeus) Lindley

Calopogon pulchellus (Salisbury) R. Brown = *Calopogon tuberosus* (Linnaeus) Britton, Sterns, & Poggenberg

Calopogon pulchellus (Salisbury) R. Brown var. *latifolius* (St. John) Fernald = *Calopogon tuberosus* (Linnaeus) Britton, Sterns, & Poggenberg

Coeloglossum bracteatum (Mühlenberg) *ex* Willdenow) Parlin = *Coeloglossum viride* (Linnaeus) Hartman var. *virescens* (Mühlenberg) Luer

Coeloglossum viride subsp. *bracteatum* (Mühlenberg ex Willdenow) Hultén = *Coeloglossum viride* (Linnaeus) Hartman var. *virescens* (Mühlenberg) Luer

Corallorhiza bigelovii S. Watson = *Corallorhiza striata* Lindley var. *vreelandii* (Rydberg) L.O. Williams

Corallorhiza corallorhiza (Linnaeus) Karsten = *Corallorhiza trifida* Chatelain

Corallorhiza maculata subsp. *occidentalis* (Lindley) Cockerell = *Corallorhiza maculata* (Rafinesque) Rafinesque var. *occidentalis* (Lindley) Ames

Corallorhiza multiflora Nuttall = *Corallorhiza maculata* (Rafinesque) Rafinesque

Corallorhiza pringlei Greenman = *Corallorhiza odontorhiza* (Willdenow) Poiret var. *pringlei* (Greenman) Freudenstein

Cypripedium calceolus Linnaeus var. *planipetalum* (Fernald) Victorin & Rousseau = *Cypripedium parviflorum* Salisbury var. *pubescens* (Willdenow) Knight

Cypripedium calceolus Linnaeus ≠ *Cypripedium parviflorum* Salisbury var. *pubescens* (Willdenow) Knight

Cypripedium calceolus Linnaeus var. *parviflorum* (Salisbury) Fernald = *Cypripedium parviflorum* Salisbury var. *makasin* (Farwell) Sheviak

Cypripedium calceolus Linnaeus var. *parviflorum* (Salisbury) Fernald p.p. = *Cypripedium parviflorum* Salisbury var. *parviflorum*

Cypripedium calceolus Linnaeus var. *parviflorum* Salisbury p.p. = *Cypripedium parviflorum* Salisbury var. *makasin* (Farwell) Sheviak

Cypripedium calceolus Linnaeus var. *planipetalum* (Fernald) Victorin & Rousseau = *Cypripedium parviflorum* Salisbury var. *pubescens* (Willdenow) Knight

Cypripedium calceolus Linnaeus var. *pubescens* (Willdenow) Correll = *Cypripedium parviflorum* Salisbury var. *pubescens* (Willdenow) Knight

Cypripedium flavescens de Candolle = *Cypripedium parviflorum* Salisbury var. *pubescens* (Willdenow) Knight

Cypripedium daultonii V. Soukup *nom. nud.* = *Cypripedium kentuckiense* C.F. Reed

Cypripedium parviflorum var. *planipetalum* Fernald = *Cypripedium parviflorum* Salisbury var. *pubescens* (Willdenow) Knight

Cypripedium pubescens var. *makasin* Farwell = *Cypripedium parviflorum* Salisbury var. *makasin* (Farwell) Sheviak

Cypripedium pubescens Willdenow = *Cypripedium parviflorum* Salisbury var. *pubescens* (Willdenow) Knight

Cypripedium spectabile Salisbury = *Cypripedium reginae* Walter

Cypripedium veganum Cockerell & Barber = *Cypripedium parviflorum* Salisbury var. *pubescens* (Willdenow) Knight

Cytherea bulbosa (Linnaeus) House = *Calypso bulbosa* (Linnaeus) Oakes var. *americana* (R. Brown) Luer

Dactylorhiza viridis (Linnaeus) R.M. Bateman, A. Pridgeon, & M.W. Chase = *Coeloglossum viride* (Linnaeus) Hartman

Epipactis latifolia (Linnaeus) Allioni = *Epipactis helleborine* (Linnaeus) Cranz

Fissipes acaulis (Aiton) Small = *Cypripedium acaule* Aiton

Galeorchis spectabilis (Linnaeus) Rydberg = *Galearis spectabilis* (Linnaeus) Rafinesque

Goodyera decipiens (Hooker) F.T. Hubbard = *Goodyera oblongifolia* Rafinesque

Gyrostachys stricta Rydberg = *Spiranthes romanzoffiana* Chamisso

Habenaria ×media (Rydberg) Niles = *Platanthera huronensis* (Nuttall) Lindley

Habenaria blephariglottis (Willdenow) Hooker = *Platanthera blephariglottis* (Willdenow) Lindley

Habenaria blephariglottis (Willdenow) Hooker = *Platanthera blephariglottis* (Willdenow) Lindley var. *blephariglottis*

Habenaria borealis Chamisso var. *albiflora* Chamisso = *Platanthera dilatata* (Pursh) Lindley var. *albiflora* (Chamisso) Ledebour

Habenaria borealis Chamisso var. *viridiflora* Chamisso = *Platanthera stricta* Lindley

Habenaria bracteata (Mühlenberg *ex* Willdenow) R. Brown *ex* Aiton *f.* = *Coeloglossum viride* (Linnaeus) Hartman var. *virescens* (Mühlenberg) Luer

Habenaria ciliaris (Linnaeus) R. Brown = *Platanthera ciliaris* (Linnaeus) Lindley

Habenaria clavellata (Michaux) Sprengel = *Gymnadeniopsis clavellata* (Michaux) Rydberg

Habenaria clavellata var. *ophioglossoides* Fernald = *Gymnadeniopsis clavellata* (Michaux) Rydberg var. *ophioglossoides* (Fernald) W.J. Schrenk

Habenaria cristata (Michaux) R. Brown = *Platanthera cristata* (Michaux) Lindley

Habenaria dilatata (Pursh) Hooker var. *albiflora* (Chamisso) Correll = *Platanthera dilatata* (Pursh) Lindley var. *albiflora* (Chamisso) Ledebour

Habenaria flava (Linnaeus) R. Brown *ex* Sprengel = *Platanthera flava* (Linnaeus) Lindley var. *flava*

Habenaria flava var. *herbiola* (R. Brown *ex* Aiton) Ames & Correll = *Platanthera flava* (Linnaeus) Lindley var. *herbiola* (R. Brown) Luer

Habenaria flava var. *virescens sensu* Fernald = *Platanthera flava* (Linnaeus) Lindley var. *herbiola* (R. Brown) Luer

Habenaria herbiola R. Brown *ex* Aiton = *Platanthera flava* (Linnaeus) Lindley var. *herbiola* (R. Brown) Luer
Habenaria hookeri Torrey = *Platanthera hookeri* (Torrey) Lindley
Habenaria hyperborea (Linnaeus) R. Brown *ex* Aiton ≠ *Platanthera aquilonis* Sheviak
Habenaria hyperborea (Linnaeus) R. Brown var. *huronensis* (Nuttall) Farwell = *Platanthera huronensis* (Nuttall) Lindley
Habenaria lacera (Michaux) R. Brown = *Platanthera lacera* (Michaux) G. Don
Habenaria leucophaea (Nuttall) A. Gray = *Platanthera leucophaea* (Nuttall) Lindley
Habenaria leucophaea (Nuttall) A. Gray var. *praeclara* (Sheviak & Bowles) Cronquist = *Platanthera praeclara* Sheviak & Bowles
Habenaria nivea (Nuttall) Sprengel = *Gymnadeniopsis nivea* Nuttall
Habenaria orbiculata (Pursh) Torrey = *Platanthera orbiculata* (Pursh) Lindley
Habenaria peramoena A. Gray = *Platanthera peramoena* (A. Gray) A. Gray
Habenaria psycodes (Linnaeus) Sprengel = *Platanthera psycodes* (Linnaeus) Lindley
Habenaria saccata Greene = *Platanthera stricta* Lindley
Habenaria unalascensis (Sprengel) S. Watson = *Piperia unalascensis* (Sprengel) Rydberg
Habenaria viridis (Linnaeus) R. Brown *ex* Aiton f. = *Coeloglossum viride* (Linnaeus) Hartman var. *viride*
Habenaria viridis Linnaeus var. *bracteata* (Mühlenberg *ex* Willdenow) Reichenbach *ex* Gray = *Coeloglossum viride* (Linnaeus) Hartman var. *virescens* (Mühlenberg) Luer
Helleborine gigantea (Douglas) Druce = *Epipactis gigantea* Douglas *ex* Hooker
Ibidium beckii (Lindley) House = *Spiranthes lacera* Rafinesque var. *gracilis* (Bigelow) Luer
Ibidium beckii House = *Spiranthes tuberosa* Rafinesque
Ibidium cernuum (Linnaeus) House = *Spiranthes cernua* (Linnaeus) L.C. Richard
Ibidium gracile (Bigelow) House = *Spiranthes lacera* (Rafinesque) Rafinesque var. *gracilis* (Bigelow) Luer
Ibidium laciniatum (Small) House = *Spiranthes laciniata* (Small) Ames
Ibidium odoratum (Nuttall) House = *Spiranthes odorata* (Nuttall) Lindley
Ibidium plantagineum (Rafinesque) House = *Spiranthes lucida* (H.H. Eaton) Ames
Ibidium strictum (Rydberg) House = *Spiranthes romanzoffiana* Chamisso
Ibidium vernale (Engelmann & A. Gray) House = *Spiranthes vernalis* Engelmann & A. Gray
Isotria affinis (Austin *ex* A. Gray) Rydberg = *Isotria medeoloides* (Pursh) Rafinesque

Limnorchis dilatata (Pursh) Rydberg *ex* Britton = *Platanthera dilatata* (Pursh) Lindley
Limnorchis media Rydberg = *Platanthera huronensis* (Nuttall) Lindley
Limnorchis stricta (Lindley) Rydberg = *Platanthera stricta* Lindley
Limodorum pulchellum Salisbury = *Calopogon tuberosus* (Linnaeus) Britton, Sterns, & Poggenberg
Limodorum tuberosum Linnaeus = *Calopogon tuberosus* (Linnaeus) Britton, Sterns, & Poggenberg
Microstylis unifolia (Michaux) Britton, Sterns, & Poggenberg = *Malaxis unifolia* Michaux
Neottia australis (Lindley) Szlachetko = *Listera australis* Lindley
Neottia gracilis Bigelow = *Spiranthes lacera* Rafinesque var. *gracilis* (Bigelow) Luer
Neottia lacera Rafinesque = *Spiranthes lacera* Rafinesque var. *lacera*
Orchis spectabilis Linnaeus = *Galearis spectabilis* (Linnaeus) Rafinesque
Piperia dilatata (Pursh) Szlachetko & P. Rutkowski = *Platanthera dilatata* (Pursh) Lindley var. *dilatata*
Piperia dilatata (Pursh) Szlachetko & P. Rutkowski var. *albiflora* (Chamisso) Szlachetko & P. Rutkowski = *Platanthera dilatata* (Pursh) Lindley var. *albiflora* (Chamisso) Ledebour
Platanthera ×*media* (Rydberg) Luer = *Platanthera huronensis* (Nuttall) Lindley
Platanthera clavellata (Michaux) Luer = *Gymnadeniopsis clavellata* (Michaux) Rydberg ex Britton
Platanthera dilatata (Pursh) Lindley var. *chlorantha* Hultén = *Platanthera huronensis* Lindley
Platanthera foetida Geyer *ex* Hooker ≠ *Piperia unalascensis* (Sprengel) Rydberg
Platanthera gracilis Lindley = *Platanthera stricta* Lindley
Platanthera hyperborea (Linnaeus) Lindley ≠ *Platanthera aquilonis* Sheviak
Platanthera hyperborea (Linnaeus) Lindley var. *huronensis* (Nuttall) Luer = *Platanthera huronensis* (Nuttall) Lindley
Platanthera nivea (Nuttall) Luer = *Gymnadeniopsis nivea* Nuttall
Platanthera repens (Nuttall) Wood = *Habenaria repens* Nuttall
Platanthera saccata (Greene) Hultén = *Platanthera stricta* Lindley
Platanthera unalascensis (Sprengel) Kurtz = *Piperia unalascensis* (Sprengel) Rydberg
Pogonia affinis Austin *ex* A. Gray = *Isotria medeoloides* (Pursh) Rafinesque
Pogonia verticillata (Mühlenberg *ex* Willdenow) Nuttall = *Isotria verticillata* (Mühlenberg ex Willdenow) Rafinesque
Spiranthes beckii Lindley = *Spiranthes lacera* (Rafinesque) Rafinesque var. *gracilis* (Bigelow) Luer
Spiranthes cernua (Linnaeus) Richard var. *odorata* (Nuttall) Correll = *Spiranthes odorata* (Nuttall) Lindley
Spiranthes gracilis (Bigelow) Beck = *Spiranthes lacera* (Rafinesque) Rafinesque var. *gracilis* (Bigelow) Luer

Spiranthes grayi Ames = *Spiranthes tuberosa* Rafinesque
Spiranthes intermedia Ames ≠ *Spiranthes casei* Catling & Cruise var. *casei*
Spiranthes plantaginea Rafinesque = *Spiranthes lucida* (H.H. Eaton) Ames
Spiranthes romanzoffiana Chamisso var. *diluvialis* (Sheviak) S.L. Welsh = *Spiranthes diluvialis* Sheviak
Spiranthes stricta (Rydberg) A. Nelson *ex* J.M. Coulter & A. Nelson = *Spiranthes romanzoffiana* Chamisso
Spiranthes tuberosa var. *grayi* (Ames) Fernald = *Spiranthes tuberosa* Rafinesque
Spiranthes unalascensis Sprengel = *Piperia unalascensis* (Sprengel) Rydberg
Spiranthes vernalis Engelmann & Gray ≠ *Spiranthes casei* Catling & Cruise

Using Luer

Additions, corrections, nomenclatural changes, and comments for Luer (1975), *The Native Orchids of the United States and Canada excluding Florida*, pertaining to *Wild Orchids of the Prairies and Great Plains Region of North America*

For those fortunate enough to own or have access to a copy of Carlyle Luer's original works on the orchids of the United States and Canada, the following additions, corrections, and comments are assembled. These in no way should detract from the usefulness of those books; they simply allow for more than twenty-five years of research and nomenclatural changes as well as for the addition of several species that had not been described as of the date of publication. Names of authors may be found in the text and checklist of this book. No attempt has been made to completely rework the keys or the index.

Introduction

p. 12 for *Cypripedium calceolus* var. *planipetalum* read *Cypripedium parviflorum* var. *pubescens* extreme form; *Calopogon tuberosus* var. *latifolius* and var. *nanus* read dwarf forms; *P. hookeri* var. *abbreviata* read *P. hookeri* forma *abbreviata*; *P. obtusata* var. *collectanea* read *P. obtusata* forma *collectanea*; *P. orbiculata* var. *lehorsii* read *P. orbiculata* forma *lehorsii*; for *Cypripedium calceolus* var. *pubescens* read *Cypripedium parviflorum* var. *pubescens*; for *Malaxis monophyllos* var. *brachypoda* read *Malaxis brachypoda*; for *Platanthera hyperborea* read *Platanthera aquilonis*

p. 39, couplets 7 and 8, see *Cypripedium* key in text on page 53

pp. 40, 41, pl. 1:3,4 forma *albiflorum*

pp. 42, 43, pl. 2:1 forma *albiflorum*

p. 44 for *Cypripedium calceolus* Linnaeus var. *pubescens* (Willdenow) Correll read *Cypripedium parviflorum* Salisbury var. *pubescens* (Willdenow) Knight

pp. 44, 45, pl. 3, p. 46, 47, pl. 4 for *Cypripedium calceolus* var. *pubescens* read *Cypripedium parviflorum* var. *pubescens*

p. 48 for *Cypripedium calceolus* Linnaeus var. *parviflorum* Salisbury read *Cypripedium parviflorum* Salisbury var. *makasin* (Farwell) Sheviak

p. 49 for *Cypripedium* ×*andrewsii* Fuller read *Cypripedium* ×*andrewsii* Fuller nm *andrewsii*; for *Cypripedium* ×*favillianum* Curtis read *C.* ×*andrewsii* Fuller nm *favillianum* (Curtis) Boivin; *C. calceolus* var. *parviflorum* read *Cypripedium parviflorum* var. *makasin*; for *C.* ×*landonii* read *C.* ×*andrewsii* nm *landonii* pp. 50, 51, pl. 5:1,2 for *Cypripedium calceolus* var. *parviflorum* read *Cypripedium parviflorum* var. *makasin*; pl. 5:3,4,5 for *Cypripedium calceolus* Linnaeus var. *planipetalum* (Fernald) Victorin & Rousseau read *C. parviflorum* var. *pubescens* extreme expression

p. 52, 53, pl. 6:5 read *Cypripedium* ×*andrewsii* nm. *favillianum*; pl. 6:6 read *Cypripedium* ×*andrewsii* nm. *andrewsii*

pp. 56, 57, pl. 8:5 for (*Cypripedium album*) read forma *albiflorum*

pp. 76, 77, pl. 15:3 forma *viridens*

pp. 82, 83, pl. 17:5 forma *viridis*

pp. 84, 85, pl. 18:4 forma *viridens*

p. 99 couplet l0a for *S. intermedia* read *S. casei*

p. 100 couplet 20 for *S. intermedia* read *S. casei*

p. l05, pl. 22:5 forma *albolabia*

p. 106 for *Spiranthes intermedia* read *Spiranthes casei* var. *casei*

p. 107, pl. 23:3–5 *Spiranthes casei* var. *casei*

p. 108 for *Spiranthes intermedia* Ames read *Spiranthes casei* Catling & Cruise var. *casei*

p. 114 read var. *ovalis*

pp. 142, 143, pl. 34:5 forma *reticulata*

pp. l44, 145, pl. 35:7 forma *ophioides*

pp. 148, 149, pl. 36:4 forma *willeyi*; 5 forma *gordinierii*; for *Cypripedium calceolus* var. *pubescens* read *Cypripedium parviflorum* var. *pubescens*

p. 178 couplet 13a for *P. lacera* var. *terrae-novae* read *P.* ×*andrewsii*; couplet 22 for *P. orbiculata* var. *orbiculata* read *P. orbiculata*

p. 179 couplet 36 for *P. hyperborea* var. *hyperborea* read *P. aquilonis*; couplet 36a for *P. hyperborea* var. *gracilis* read *P. huronensis*; couplet 37 for *P. hyperborea* var. *huronensis* read *P. huronensis*; couplet 37a for *P. hyperborea* var. *viridiflora* read *P. huronensis*; couplet 34a for *P.* ×*media* read *P. huronensis*; couplet 39 for *P. hyperborea* var. *purpurascens* read *P. purpurascens*

pp. 192, 193, pl. 48:8,9 for *Platanthera lacera* var. *terrae-novae* read *P.* ×*andrewsii*

p. 194 for *Platanthera lacera* (Michaux) Don var. *terrae-novae* (Fernald) Luer read *P.* ×*andrewsii* (White ex Niles) Luer

pp. 196, 197, pl. 49:8 forma *albiflora*

p. 204 for *Platanthera nivea* (Nuttall) Luer read *Gymnadeniopsis nivea* (Nuttall) Rydberg

p. 206 for *Platanthera clavellata* (Michaux) Luer read *Gymnadeniopsis clavellata* (Michaux) Rydberg

p. 218 for *Platanthera hookeri* var. *abbreviata* read forma *abbreviata*

pp. 220, 221, pl. 59:1,2 for *Platanthera orbiculata* var. *orbiculata* read *Platanthera orbiculata*; 3–4 for *Platanthera orbiculata* var. *macrophylla* read *Platanthera macrophylla*

p. 222 for *Platanthera orbiculata* (Pursh) Lindley var. *macrophylla* (Goldie) Luer read *Platanthera macrophylla* (Goldie) P.M. Brown

p. 228 for *Platanthera hyperborea* (Linnaeus) Lindley var. *hyperborea* read *Platanthera aquilonis* Sheviak

p. 229 for *P.* ×*media* (Rydberg) Luer read *P. huronensis* (Nuttall) Lindley

pp. 230, 231 for *Platanthera hyperborea* (Linnaeus) Lindley var. *huronensis* read *Platanthera huronensis* (Nuttall) Lindley; pl. 61:1,3,4 read *Platanthera aquilonis*; 5–8 read *Platanthera huronensis*

p. 232 for *Platanthera hyperborea* (Linnaeus) Lindley var. *gracilis* (Lindley) Luer read *Platanthera huronensis* (Nuttall) Lindley

p. 233 for *Platanthera hyperborea* (Linnaeus) Lindley var. *viridiflora* (Chamisso) Luer read *Platanthera huronensis* (Nuttall) Lindley

pp. 234, 235, pl. 62:1–5 read *Platanthera huronensis*; 6–9 *Platanthera purpurascens*

pp. 270, 271, pl. 74:1 forma *albolabia*

pp. 318, 319, pl. 89:1,9 *Corallorhiza maculata* var. *occidentalis* forma *immaculata*

pp. 320, 321, pl. 90:3 forma *albolabia*

pp. 324, 325, pl. 92:4 forma *eburnea*

pp. 338, 339, pl. 96 forma *albiflora*

The following taxa found within the prairies and Great Plains region are not treated in *The Native Orchids of the United States and Canada excluding Florida*:

Calopogon oklahomensis
Corallorhiza maculata var. *occidentalis*
Corallorhiza odontorhiza var. *pringlei*
Cypripedium kentuckiense
Cypripedium parviflorum var. *parviflorum*
Platanthera praeclara
Spiranthes diluvialis
Spiranthes ovalis var. *erostellata*
Spiranthes sylvatica

Cryptic Species, Species Pairs, and Varietal Pairs

When two species appear to be very closely related and have a similar morphology they are often referred to as species pairs. Although not necessarily a scientific or taxonomic term, the designation is often helpful in recognizing two species that are often difficult to determine, whether in the field or in the herbarium. Their taxonomic history usually involves synonyms and/or recognition at different taxonomic levels, i.e. subspecies, varieties, or rarely forma. Validation of the two taxa at species level usually involves studies of pollinators, habit, habitat, range, morphology, and, in more recent years, DNA analyses. Few true species pairs exist for the prairies and Great Plains region of North America but several varietal pairs are found in this area. Fortunately a few simple morphological characters easily separate the taxa. In each case these are characters that can be found within the key to the species. Also, range and habitat are often well separated.

Corallorhiza maculata var. *maculata*
spotted coralroot
Corallorhiza maculata var. *occidentalis*
western spotted coralroot
Differ in the shape of the lip and degree of spreading of the tepals; ranges overlap but var. *occidentalis* usually flowers earlier than var. *maculata*.

var. *maculata*   var. *occidentalis*

Corallorhiza odontorhiza var. *odontorhiza*
autumn coralroot
Corallorhiza odontorhiza var. *pringlei*
Pringle's autumn coralroot
Differ in shape of lip and openness of flower—i.e., cleistogamous vs. chasmogamous; also, range very limited for var. *pringlei*.

var. *odontorhiza*   var. *pringlei*

Corallorhiza striata var. *striata*
striped coralroot
Corallorhiza striata var. *vreelandii*
Vreeland's striped coralroot
Differ in color, shape, and size of flowers; ranges overlap in western North America.

var. *striata*   var. *vreelandii*

Cypripedium parviflorum var. *parviflorum*
southern small yellow lady's-slipper
Cypripedium parviflorum var. *makasin*
northern small yellow lady's-slipper
Cypripedium parviflorum var. *pubescens*
large yellow lady's-slipper
Range differences, with var. *parviflorum* basically southeastern, var. *makasin* northern, and var. *pubescens* throughout much of United States and Canada; subtle differences in coloration of petals and sepals, shape of lip, fragrance, and habitat.

var. *parviflorum*   var. *makasin*   var. *pubescens*

Platanthera aquilonis
northern green bog orchis
Platanthera huronensis
green bog orchis
Differ in lip shape and color, pollination biology, and somewhat in habit and habitat; ranges overlap throughout.

P. aquilonis   *P. huronensis*

Platanthera leucophaea
eastern prairie fringed orchis
Platanthera praeclara
western prairie fringed orchis
Clearly separated by floral characters, pollinators, range, and size; classic species pair.

P. *leucophaea* P. *praeclara*

Spiranthes magnicamporum
Great Plains ladies'-tresses
Spiranthes diluvialis
Ute ladies'-tresses
Spiranthes romanzoffiana
hooded ladies'-tresses
Differ in floral biology, i.e., lip shape, petal and sepal positioning, and flower position; *Spiranthes diluvialis*, is the amphidiploid progeny of *S. magnicamporum* and *S. romanzoffiana*, with morphological characters intermediate with the parents.

S. *magnicamporum* S. *diluvialis* S. *romanzoffiana*

Spiranthes lacera var. *lacera*
northern slender ladies'-tresses
Spiranthes lacera var. *gracilis*
southern slender ladies'-tresses
These two very similar varieties often overlap their ranges in the central states, both having disjunct sites north and south of their respective ranges. The well-spaced-out flowers on var. *lacera* contrast with the crowded flowers of var. *gracilis*. At flowering time the leaves are usually present in var. *lacera* and usually absent in var. *gracilis*. The lip on var. *gracilis* is broader and crisper with a more distinctive green throat, and that of var. *lacera* is narrower with the green throat less prominent. The sepals on var. *lacera* are approximate to appressed whereas on var. *gracilis* they are nearly spreading, giving the flower a larger overall appearance. See descriptions on pages 182 and 184 for details.

var. *lacera*   var. *gracilis*

Part 4
~
Orchid Hunting

1. The Northern Prairie Region

The northern prairies, primarily in portions of northern Illinois, Wisconsin, Minnesota, Manitoba, Saskatchewan, North Dakota, and South Dakota, have a common orchid flora, and although some species may occur in one state and not another, several species may be found throughout this area. Many of the prairies are small and locally preserved, but in southern Manitoba the vast grasslands near Tolstoi are the largest unspoiled prairie areas in the northern Midwest.

Species that may be found in the northern prairies region include:
Calopogon oklahomensis, Oklahoma grass-pink
Calopogon tuberosus, common grass-pink
Cypripedium candidum, small white lady's-slipper
Cypripedium parviflorum var. *makasin*, northern small yellow lady's-slipper
Cypripedium parviflorum var. *pubescens*, large yellow lady's-slipper
Cypripedium reginae, showy lady's-slipper
Gymnadeniopsis clavellata, little club-spur orchis
Liparis loeselii, Loesel's twayblade
Malaxis unifolia, green adder's-mouth
Platanthera aquilonis, northern green bog orchis
Platanthera flava var. *herbiola*, northern tubercled orchis
Platanthera huronensis, green bog orchis
Platanthera lacera, green fringed orchis
Platanthera leucophaea, eastern prairie fringed orchis
Platanthera praeclara, western prairie fringed orchis
Platanthera psycodes, small purple fringed orchis

Pogonia ophioglossoides, rose pogonia
Spiranthes cernua, nodding ladies'-tresses
Spiranthes lacera var. *lacera*, northern slender ladies'-tresses
Spiranthes lacera var. *gracilis*, southern slender ladies'-tresses
Spiranthes magnicamporum, Great Plains ladies'-tresses
Spiranthes romanzoffiana, hooded ladies'-tresses

A visit to the prairies of Minnesota, Wisconsin, and adjacent northern Illinois in May will begin the prairie orchid season with searches for plants of *Cypripedium candidum*, the small white lady's-slipper, and watchful eyes may find some of the hybrids *C. candidum* produces with *C. parviflorum* var. *makasin*, the northern small yellow lady's-slipper, and *C. parviflorum* var. *pubescens*, the large yellow lady's-slipper: *C.* ×*andrewsii* and *C.* ×*andrewsii* nm *favillianum* respectively. Although harder to detect, *Cypripedium* ×*andrewsii* nm *landonii*, a documented backcross, may also be present. Look for the yellow-flowered species in the copses, hedgerows, and shrubby areas that occur within the prairies.

Historically *Calopogon oklahomensis*, the recently described Oklahoma grass-pink, was known from some of the more southern prairies in Minnesota and Wisconsin. Early flowering grass-pinks found in these areas should be very carefully examined to determine their identification and simply not assumed to be early plants of the common grass-pink, *Calopogon tuberosus*.

Late June into July is one of the most beautiful times within the northern and eastern prairies, with the largest array of prairie wildflowers in full bloom. Compass plant, wild indigo, Indian paintbrushes, coreopsis, milkweeds, and lilies all contribute to this tapestry. Among this riot of color, in suitable areas, especially some of the moister prairies, may be found *Pogonia ophioglossoides*, rose pogonia, *Calopogon tuberosus*, the common grass-pink, *Platanthera flava* var. *herbiola*, the northern tubercled orchis, and, in the few areas that appear to suit its whims, *P. leucophaea*, the eastern prairie fringed orchis. The riverine prairies in Illinois Beach State Park and Chiwaukee Prairie—only a few miles apart in Zion, Illinois, and nearby Wisconsin—are the poster boys for this early summer display. The capricious nature of the eastern prairie fringed orchis is such that it is rarely reliable in consecutive years and may put on a dazzling display one year and then produce no flowering plants the next. Although a few leaves may be produced many years even those are absent as the plants lie dormant.

In western Minnesota and southern Manitoba *Platanthera praeclara*, the western prairie fringed orchis, is the dominant prairie orchid to be found. More reliable in its flowering habits than its cousin, its giant stems and massive flowers proudly hold their heads among the grasses and colorful herbs. Roadside ditches or wetter depressions within the prairies often support colonies of *P. aquilonis*, the northern green bog orchis; *P. huronensis*, the green bog orchis; *Liparis loeselii*, Loesel's twayblade; *Malaxis unifolia*, the green adder's-mouth; *Gymnadeniopsis clavellata*, the little club-spur orchis; and *Calopogon tuberosus*, the common grass-pink. In areas where the prairie regions are near the border of the northern woodlands also look for *Cypripedium reginae*, the showy lady's-slipper. If it is present you will have no difficulty in spotting it! Other species that may occur in the northern prairies and nearby areas might include, in the north, *Spiranthes lacera* var. *lacera*, the northern slender ladies'-tresses and, in the south, *S. lacera* var. *gracilis*, the southern slender ladies'-tresses, and *Platanthera lacera*, the green fringed orchis. In extreme southeastern Manitoba *P. psycodes*, the small purple fringed orchis, has been found and the possibility of the hybrid, *P. ×andrewsii*, Andrew's hybrid fringed orchis, is increased. It is here the small purple fringed orchis reaches the northwestern limit of its range.

Autumn has few orchids in this area but the myriad of asters and goldenrods make for a pleasant tapestry. The typical prairies species, *Spiranthes magnicamporum*, the Great Plains ladies'-tresses, is found in all of these states and provinces, although widely scattered. *Spiranthes romanzoffiana*, the hooded ladies'-tresses, is widespread throughout all of the northern states and not confined to the open grasslands.

South of the Canadian provinces *Spiranthes cernua*, the nodding ladies'-tresses, may also be seen. Plants may need to be carefully examined to determine a correct identification. Late flowering slender ladies'-tresses, *S. lacera*, in either variety may also be found.

2. Central Mississippi River Valley Region

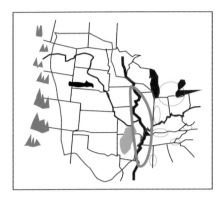

What were once wide-open prairie lands along the Mississippi River from central Illinois to Arkansas have undergone both agricultural and commercial development in the past 150 years. Small remnant prairies have survived, usually as a result of the efforts of local conservation groups, and it is in these prairies and adjacent woodlands that we find so many orchids.

Species that may be found in the central Mississippi River Valley region include:
  Aplectrum hyemale, Adam-and-Eve
  Calopogon oklahomensis, Oklahoma grass-pink
  Calopogon tuberosus, common grass-pink
  Corallorhiza maculata var. *maculata*, spotted coralroot
  Corallorhiza odontorhiza var. *odontorhiza*, autumn coralroot
  Corallorhiza wisteriana, Wister's coralroot
  Cypripedium parviflorum var. *parviflorum*, southern small yellow lady's-slipper
  Cypripedium parviflorum var. *pubescens*, large yellow lady's-slipper
  Galearis spectabilis, showy orchis
  Goodyera pubescens, downy rattlesnake orchis
  Gymnadeniopsis clavellata, little club-spur orchis
  Habenaria repens, water-spider orchis
  Hexalectris spicata, crested coralroot
  Isotria verticillata, large whorled pogonia
  Liparis liliifolia, lily-leaved twayblade
  Liparis loeselii, Loesel's twayblade
  Listera australis, southern twayblade

Malaxis unifolia, green adder's-mouth
Platanthera ciliaris, orange fringed orchis
Platanthera cristata, orange crested orchis
Platanthera flava var. *flava*, southern tubercled orchis
Platanthera flava var. *herbiola*, northern tubercled orchis
Platanthera lacera, green fringed orchis
Platanthera leucophaea, eastern prairie fringed orchis
Platanthera peramoena, purple fringeless orchis
Platanthera praeclara, western prairie fringed orchis
Pogonia ophioglossoides, rose pogonia
Spiranthes cernua, nodding ladies'-tresses
Spiranthes lacera var. *gracilis*, southern slender ladies'-tresses
Spiranthes laciniata, lace-lipped ladies'-tresses
Spiranthes magnicamporum, Great Plains ladies'-tresses
Spiranthes odorata, fragrant ladies'-tresses
Spiranthes ovalis var. *erostellata*, northern oval ladies'-tresses
Spiranthes sylvatica, woodland ladies'-tresses
Spiranthes tuberosa, little ladies'-tresses
Spiranthes vernalis, grass-leaved ladies'-tresses
Tipularia discolor, crane-fly orchis
Triphora trianthophora, three birds orchis

Spring in the lower Mississippi River Valley in northern Louisiana and eastern Arkansas may feature only one of the prairie orchids in the postage-stamp prairies that dot the rice fields, but the presence of *Calopogon oklahomensis*, the Oklahoma grass-pink, more than makes up for the lack of other orchids. In a small strip prairie in eastern Arkansas, bordered by railroad tracks and acres of rice fields that once were prairies, the pink, lavender, magenta, and white flowers of the Oklahoma grass-pink are surrounded by dozens of spring wildflowers weaving a tapestry of color. Indian paintbrushes, spiderwort, coreopsis, indigo, milkweeds, and more only add to the early May display. It is among these larger plants that the delicate stems of the Oklahoma grass-pink are nestled. They are not visible at first from a distance, but once you venture into the prairies they reveal themselves, often by the hundreds!

Northward in the valley to Missouri and Illinois the remnant prairies continue the spring array, but here most records of the Oklahoma grass-pink are historical, as are those of the small white ladies'-slipper, *Cypripedium candidum*. Northern Illinois and eastern Iowa do have remaining populations of the small white lady's-slipper, and in the nearby groves and woodlands plants of the yellow lady's-slipper, *C. parviflorum* in variety, may continue to thrive. It is here that the hybrids are most easily found.

By early summer *Platanthera leucophaea*, the eastern prairie fringed orchis, found primarily east of the Mississippi River, and *Platanthera praeclara*, the western prairie fringed orchis, are starting to display their midsummer array. In those few remaining spots a myriad of wildflowers may be scattered with a few tall, showy, flowering stems of the prairie fringed orchises. As mentioned previously, plants of both prairie fringed orchises are not always reliable from year to year and they both respond favorably to controlled burns. *Spiranthes vernalis*, the grass-leaved ladies'-tresses, *S. tuberosa*, the little ladies'-tresses, and *S. sylvatica*, the woodland ladies'-tresses (in Arkansas and Louisiana), are all late spring or summer flowering species that may be found in these areas or along bordering woodlands. *Spiranthes tuberosa*, the little ladies'-tresses, is especially fond of old, thin-soil cemeteries and not unexpectedly accompanied by *S. lacera* var. *gracilis*, the southern slender ladies'-tresses.

Depending on your locale either *Platanthera flava* var. *herbiola*, the northern tubercled orchis, or *P. flava* var. *flava*, the southern tubercled orchis, may occasionally be seen in the damp meadow-like prairies. Additional species of the summer orchids that are just as happily at home in a variety of wetland and meadow habitats but also frequent selected prairies include *Calopogon tuberosus*, common grass-pink; *Habenaria repens*, the water-spider orchis (southern Arkansas and Louisiana); *Liparis loeselii*, Loesel's twayblade; *Listera australis*, the southern twayblade; *Malaxis unifolia*, the green adder's-mouth; *Platanthera ciliaris*, the orange fringed orchis; *P. cristata*, the orange crested orchis (southern Arkansas and Louisiana); *P. lacera*, the green fringed orchis; *P. peramoena*, the purple fringeless orchis, and *Pogonia ophioglossoides*, the rose pogonia.

Late summer and autumn bring another selection of ladies'-tresses as well. The giant of the ladies'-tresses of this region, *Spiranthes magnicamporum*, the Great Plains ladies'-tresses, is found both in open, flat grasslands and among the hill and bluff prairies. This species has a distinct preference for soils that are decidedly alkaline. *Spiranthes cernua*, the nodding ladies'-tresses, is more cosmopolitan in its habitat preferences and may be found in several local races throughout the entire Mississippi Valley region.

In southern Arkansas and northern Louisiana *Spiranthes odorata*, the fragrant ladies'-tresses, will be flowering well into autumn. Although it is equally at home in rich, shaded river bottoms, plants can easily be found in the wet prairies or drainage ditches adjacent to the prairies.

Many of the rich woodlands that border the prairies of the Mississippi Valley have another set of equally interesting orchids. These are woodland species that range, for the most part, only as far west as this area. Although a few may venture into the open, they certainly are not in the grasslands but may be seen not far from typical prairies in many states. They include *Aplectrum hyemale*, Adam-and-Eve; *Corallorhiza maculata* var. *maculata*, the spotted coralroot; *C. odontorhiza* var. *odontorhiza*, the autumn coralroot; *C. wisteriana*, Wister's coralroot; *Cypripedium parviflorum* var. *pubescens*, the large yellow lady's-slipper; *Galearis spectabilis*, the showy orchis; *Goodyera pubescens*, the downy rattlesnake orchis; *Gymnadeniopsis clavellata*, the little club-spur orchis; *Isotria verticillata*, the large whorled pogonia; *Liparis liliifolia*, the lily-leaved twayblade; *Platanthera peramoena*, the purple fringeless orchis; *Spiranthes ovalis* var. *erostellata*, the northern oval ladies'-tresses; *Tipularia discolor*, the crane-fly orchis, and *Triphora trianthophora*, three birds orchis.

3. Islands in the Plains

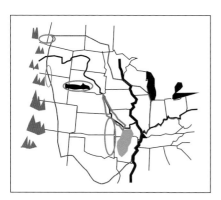

Throughout the Great Plains exist many isolated wooded areas. These woodlands may be stands along river systems, independent copses that have developed around springs, or the result of early towns and development. Whatever their origin they all support several species of orchids that otherwise might not be seen within the plains. The combination of added shelter, moisture, and often somewhat different soils all contribute to the habitat. Some of the most notable areas are along the Missouri River, particularly in Nebraska and Missouri, and the hill country in eastern Kansas and eastern Oklahoma.

Although several species are restricted to either the northern areas or the southern areas, species that may be found in these islands in the plains include:

Aplectrum hyemale, Adam-and-Eve
Coeloglossum viride var. *virescens*, long-bracted green orchis
Corallorhiza maculata var. *maculata*, spotted coralroot
Corallorhiza maculata var. *occidentalis*, western spotted coralroot
Corallorhiza striata var. *striata*, striped coralroot
Corallorhiza striata var. *vreelandii*, Vreeland's striped coralroot
Corallorhiza trifida, early coralroot
Corallorhiza wisteriana, Wister's coralroot
Cypripedium parviflorum var. *parviflorum*, southern small yellow lady's-slipper
Cypripedium parviflorum var. *pubescens*, large yellow lady's-slipper
Epipactis gigantea, stream orchid
Galearis spectabilis, showy orchis

Goodyera oblongifolia, giant rattlesnake orchis
Hexalectris spicata, crested coralroot
Platanthera oiliaris, orange fringed orchis
Platanthera huronensis, green bog orchis
Spiranthes cernua, nodding ladies'-tresses
Spiranthes lacera var. *lacera*, northern slender ladies'-tresses
Spiranthes lacera var. *gracilis*, southern slender ladies'-tresses
Spiranthes lucida, shining ladies'-tresses
Spiranthes magnicamporum, Great Plains ladies'-tresses
Spiranthes ovalis var. *erostellata*, northern oval ladies'-tresses
Spiranthes romanzoffiana, hooded ladies'-tresses
Spiranthes vernalis, grass-leaved ladies'-tresses

The gently rolling hills and valleys of eastern Kansas and especially Oklahoma present a surprising array of wild orchids in what are generally perceived as vast acres of endless plains and wheat fields. Larry Magrath (1971, 2001) has written about both of these states, and the distribution maps in his latest work on the orchids of Oklahoma show many of these species clustered in the southeastern quarter of the state. Spring comes early to the southern plains and one of the first orchids seen is usually *Corallorhiza wisteriana*, Wister's coralroot. It grows in a variety of habitats and colonies vary from year to year in their thriftiness. *Galearis spectabilis*, the showy orchis, and *Aplectrum hyemale*, the putty-root or Adam-and-Eve, can be found in rich wooded pockets, both reaching the western limit of their range. *Cypripedium parviflorum* var. *parviflorum*, the southern small yellow lady's-slipper, is scattered about, but watch carefully for plants of its larger cousin *C. parviflorum* var. *pubescens*, the large yellow lady's-slipper, especially as you venture northward. The southern small yellow lady's-slipper in eastern Oklahoma is especially interesting; in 1989 Larry Magrath and Jim Norman described a white-flowered form that has rarely been found elsewhere (Magrath, 1989).

Summer orchids in these same two areas often include the showy *Platanthera ciliaris*, the orange fringed orchis; *Hexalectris spicata*, the crested coralroot, and in Oklahoma, *Epipactis gigantea*, the stream orchid or chatterbox. Late summer and into autumn, when the plains are ablaze with fall color and the grassland ladies'-tresses are flowering, we may find here in these wooded islands a few plants of *Spiranthes ovalis* var. *erostellata*, the northern oval ladies'-tresses, and the elusive autumn coralroot, *Corallorhiza odontorhiza*.

Further north as we follow the Missouri River from Missouri through Nebraska, South Dakota, and westward, woodlands have developed along the rich banks and it is in these habitats that one best searches for wild orchids. *Cypripedium parviflorum* var. *pubescens*, the large yellow lady's-slipper, is still to be found in open aspen woodlands. It may be joined by *Coeloglossum viride* var. *virescens*, the long-bracted green orchis, and *Corallorhiza maculata* var. *maculata*, the spotted coralroot. In the northern portions of the riparian woodlands *C. striata* var. *striata*, the striped coralroot, and its rarer western sibling *C. striata* var. *vreelandii*, Vreeland's striped coralroot, may occasionally be found. Keep your eyes tuned for the usually plain green rosettes of *Goodyera oblongifolia*, the giant rattlesnake orchis, in the denser woodlands as you near Wyoming and Montana. All of the riverine habitats offer possibilities of additional discoveries as they are the prime areas for richer soils and sheltered niches.

4. The Black Hills

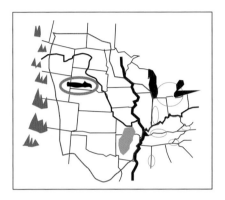

Neatly sprawled in southern South Dakota and adjacent eastern Wyoming, the Black Hills are a large, rambling range surrounded by prairies and plains. Famed for recreational facilities, a national forest, and urban areas around Sioux Falls, the Black Hills are also home to many orchid species not easily found elsewhere in the prairies and Great Plains region. The typical grassland species are in short supply, but many orchids more familiar from the northern and western areas of the United States and Canada are the features. (Note: the Black Hills of South Dakota and eastern Wyoming should not be confused with the "black hills" of central Wyoming. See page 160 for more details.)

Species that may be found in the Black Hills:
Calypso bulbosa var. *americana*, eastern fairy-slipper
Coeloglossum viride var. *virescens*, long-bracted green orchis
Corallorhiza maculata var. *maculata*, spotted coralroot
Corallorhiza maculata var. *occidentalis*, western spotted coralroot
Corallorhiza odontorhiza var. *odontorhiza*, autumn coralroot
Corallorhiza striata var. *striata*, striped coralroot
Corallorhiza striata var. *vreelandii*, Vreeland's striped coralroot
Corallorhiza trifida, early coralroot
Cypripedium parviflorum var. *makasin*, northern small yellow lady's-slipper
Cypripedium parviflorum var. *pubescens*, large yellow lady's-slipper
Epipactis gigantea, stream orchid
Goodyera oblongifolia, giant rattlesnake orchis
Goodyera repens, lesser rattlesnake orchis

Gymnadeniopsis clavellata, little club-spur orchis
Liparis loeselii, Loesel's twayblade
Listera convallarioides, broad-lipped twayblade
Malaxis unifolia, green adder's-mouth
Piperia unalascensis, Alaskan piperia
Platanthera aquilonis, northern green bog orchis
Platanthera dilatata var. *dilatata*, tall white northern bog orchis
Platanthera dilatata var. *albiflora*, bog candles
Platanthera hookeri, Hooker's orchis
Platanthera huronensis, green bog orchis
Platanthera lacera, green fringed orchis
Platanthera orbiculata, pad-leaved orchis
Platanthera stricta, slender green bog orchis
Spiranthes cernua, nodding ladies'-tresses
Spiranthes lacera var. *lacera*, northern slender ladies'-tresses
Spiranthes lacera var. *gracilis*, southern slender ladies'-tresses
Spiranthes magnicamporum, Great Plains ladies'-tresses
Spiranthes romanzoffiana, hooded ladies'-tresses

What native orchid enthusiast who has tromped the open plains and prairies in search of wild orchids would pass up the opportunity to find so many of these northern and western beauties smack in the middle of the Great Plains? Although some of the vast valleys may be remote, much of the Black Hills is easily accessible by car with good roads traversing the region, and a few of the real specialties are found in highly popular and populated areas!

Spring begins with the captivating *Calypso bulbosa* var. *americana*, the eastern fairy-slipper, often accompanied by the delicate *Corallorhiza trifida*, the early coralroot. Nearby may be lurking plants of *Cypripedium parviflorum* var. *makasin*, the northern small yellow lady's-slipper, and/or *C. parviflorum* var. *pubescens*, the large yellow lady's-slipper. Soon will follow several of the other coralroots, all larger than the early coralroot and often more colorful. Both spotted coralroots—*Corallorhiza maculata* var. *maculata*, the spotted coralroot, and *C. maculata* var. *occidentalis*, the western spotted coralroot—are found in the Black Hills with the nominate variety flowering up to a month later than the western variety. Several color forms can usually be found of both varieties. Not to be outdone by these, the largest and showiest of the coralroots, *C. striata* var. *striata*, the striped coralroot, is scattered well throughout the hills. Watch carefully for the beautiful golden-ivory form, forma *eburnea*. Considerably rarer than var. *striata* and more delicate in its coloring, *C. striata* var. *vreelandii*, Vreeland's striped coralroot, has been sparingly found here as well.

Summer offers many rewards for exploration, not the least of which is found along some of the hot springs where the western *Epipactis gigantea*, the stream orchid or chatterbox, reaches the eastern limit of its range. Often accompanied by the southern maidenhair fern, *Adiantum capillus-veneris*, this western duo has survived the onslaught of civilization around the ever-popular hot springs and both continue to grow and increase.

Venturing into the woodlands, valleys, and streamsides may reward the orchid explorer with *Goodyera oblongifolia*, the giant rattlesnake orchis; *G. repens*, the lesser rattlesnake orchis; the very rare *Listera convallarioides*, the broad-lipped twayblade; scattered plants of *Piperia unalascensis*, the Alaskan piperia, and one of the great specialities of the hills: *Platanthera orbiculata*, the pad-leaved orchis. This distinctive species is widespread in several areas of western South Dakota and eastern Wyoming, all within the Black Hills. Plants have been documented and tracked for some years and the populations are well known.

In the open wet meadows and neighboring streamsides *Platanthera dilatata* var. *dilatata*, the tall white northern bog orchis, may be seen, and in 1972 Larry Magrath collected *P. dilatata* var. *albiflora*, bog candles, representing an eastern disjunct for this western variety. In much the same areas he also found plants of *P. stricta*, the slender bog orchis—several hundred miles from their nearest known location in central Wyoming. *Platanthera stricta* is one of the most frequently seen green-flowered bog orchises in western North America and this record is a significant extension of it range. For additional details see page 000.

Other species that are well distributed throughout much of the Black Hills include *Malaxis unifolia*, the green adder's-mouth; *Platanthera aquilonis*, the northern green bog orchis; *P. lacera*, the green fringed orchis; *Spiranthes cernua*, the nodding ladies'-tresses; *S. romanzoffiana*, the hooded ladies'-tresses; *S. lacera* var. *lacera*, the northern slender ladies'-tresses, and *Coeloglossum viride* var. *virescens*, the long-bracted green orchis.

5. Foothills of the Rockies

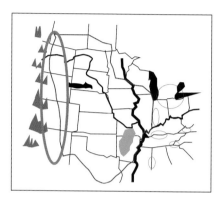

As the broad unbroken Great Plains sweep westward the foothills of the mighty Rocky Mountains begin. Such isolated ranges as the Big Horns of northern Wyoming and Montana and the Cypress Hills of Alberta and Saskatchewan are merely hints of what awaits further west. For the wild orchids of the prairies and Great Plains, these foothills offer a bit of shelter from the harsh summer suns and an opportunity for some of the western orchid species to insinuate themselves a bit further east.

Species that may be found in the foothills of the Rockies:
    Calypso bulbosa var. *americana*, eastern fairy-slipper
    Coeloglossum viride var. *virescens*, long-bracted green orchis
    Corallorhiza maculata var. *maculata*, spotted coralroot
    Corallorhiza maculata var. *occidentalis*, western spotted coralroot
    Corallorhiza striata var. *striata*, striped coralroot
    Corallorhiza striata var. *vreelandii*, Vreeland's striped coralroot
    Corallorhiza trifida, early coralroot
    Cypripedium parviflorum var. *pubescens*, large yellow lady's-slipper
    Epipactis gigantea, stream orchid
    Goodyera oblongifolia, giant rattlesnake orchis
    Goodyera repens, lesser rattlesnake orchis
    Listera convallarioides, broad-lipped twayblade
    Piperia unalascensis, Alaskan piperia
    Platanthera aquilonis, northern green bog orchis
    Platanthera dilatata var. *dilatata*, tall white northern bog orchis

Platanthera dilatata var. *albiflora*, bog candles
Platanthera praeclara, western prairie fringed orchis
Spiranthes diluvialis, Ute ladies'-tresses
Spiranthes magnicamporum, Great Plains ladies'-tresses
Spiranthes romanzoffiana, hooded ladies'-tresses

With the exception of the historical record for *Platanthera praeclara*, the western prairie fringed orchis (see p. 160), the only true prairie species within the grasslands of the foothills of the Rockies is *Spiranthes magnicamporum*, the Great Plains ladies'-tresses. Calcareous bluffs and open plains from Montana south to New Mexico are frequently suitable for this species. Otherwise, it is primarily the northern-affinity orchid species found in the Black Hills of South Dakota and adjacent Wyoming that inhabit these foothills. Exploring such areas as the Big Horns and Cypress Hills, with their many roads and trails, will reward the observer with many more locations for *Calypso bulbosa* var. *americana*, eastern fairy slipper; *Coeloglossum viride* var. *virescens*, long-bracted green orchis; *Corallorhiza maculata* var. *maculata*, the spotted coralroot; *C. maculata* var. *occidentalis*, the western spotted coralroot; *C. striata* var. *striata*, the striped coralroot; *C. trifida*, the early coralroot; *Cypripedium parviflorum* var. *pubescens*, the large yellow lady's-slipper; *Goodyera oblongifolia*, the giant rattlesnake orchis; *G. repens*, the lesser rattlesnake orchis; *Piperia unalascensis*, the Alaskan piperia; *Platanthera aquilonis*, the northern green bog orchis; *P. dilatata* var. *dilatata*, the tall white northern bog orchis; *P. dilatata* var. *albiflora*, bog candles, and *Spiranthes romanzoffiana*, hooded ladies'-tresses. Three of these are particularly rare and found in the Black Hills: *Calypso bulbosa* var. *americana*, *Piperia unalascensis*, and *Platanthera dilatata* var. *albiflora*. For a full treatment of *P. dilatata* and its three varieties, see *Wild Orchids of the Pacific Northwest and Canadian Rockies* (Brown and Folsom, 2006).

In east central Colorado and southern Wyoming *Spiranthes diluvialis*, the Ute ladies'-tresses, can be found; it also has a disjunct site in not-too-distant western Nebraska. Although more abundant in the Denver/Golden (Colorado) area and westward this amphidiploid species resulted from hybridization between *S. magnicamporum* and *S. romanzoffiana*. The Ute ladies'-tresses should be carefully sought in riverine habitats in the western Great Plains. *Spiranthes magnicamporum* is found south to north central New Mexico—well within the foothills of the Rockies (Coleman, 2002).

6. On the Edge of the Ozarks

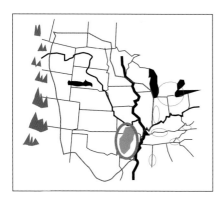

Occupying much of central Arkansas and southern Missouri, the Ozark, Boston, and Ouchita Mountains sustain an abundance of orchids. Both Summers' *Orchids of Missouri* (1996) and Slaughter's *Wild Orchids of Arkansas* (1993) give many details on those species. For our purposes in this field guide we are concerned with those that approach or encroach on the grasslands and prairies in both states. Bordering woodlands and river bluffs that occur throughout the grasslands are the best places to look for many of these upland species.

>   *Aplectrum hyemale*, Adam-and-Eve
>   *Coeloglossum viride* var. *virescens*, long-bracted green orchis
>   *Corallorhiza maculata* var. *maculata*, spotted coralroot
>   *Corallorhiza odontorhiza* var. *odontorhiza*, autumn coralroot
>   *Corallorhiza wisteriana*, Wister's coralroot
>   *Cypripedium kentuckiense*, ivory-lipped lady's-slipper
>   *Cypripedium parviflorum* var. *parviflorum*, southern small yellow lady's-slipper
>   *Cypripedium parviflorum* var. *pubescens*, large yellow lady's-slipper
>   *Galearis spectabilis*, showy orchis
>   *Goodyera pubescens*, downy rattlesnake orchis
>   *Gymnadeniopsis clavellata*, little club-spur orchis
>   *Hexalectris spicata*, crested coralroot
>   *Isotria verticillata*, large whorled pogonia
>   *Liparis liliifolia*, lily-leaved twayblade
>   *Malaxis unifolia*, green adder's-mouth
>   *Platanthera ciliaris*, orange fringed orchis

Platanthera flava var. *flava*, southern tubercled orchis
Platanthera flava var. *herbiola*, northern tubercled orchis
Platanthera lacera, green fringed orchis
Platanthera peramoena, purple fringeless orchis
Pogonia ophioglossoides, rose pogonia
Spiranthes cernua, nodding ladies'-tresses
Spiranthes lacera var. *gracilis*, southern slender ladies'-tresses
Spiranthes ovalis var. *ovalis*, southern oval ladies'-tresses
Spiranthes ovalis var. *erostellata*, northern oval ladies'-tresses
Spiranthes sylvatica, woodland ladies'-tresses
Spiranthes tuberosa, little ladies'-tresses
Spiranthes vernalis, grass-leaved ladies'-tresses
Tipularia discolor, crane-fly orchis
Triphora trianthophora, three birds orchis

In the richest of these woodlands watch in late spring for *Aplectrum hyemale*, Adam-and-Eve; *Coeloglossum viride* var. *virescens*, the long-bracted green orchis; *Cypripedium kentuckiense*, ivory-lipped lady's-slipper; *C. parviflorum* var. *parviflorum*, southern small yellow lady's-slipper; *C. parviflorum* var. *pubescens*, large yellow lady's-slipper (in the northern counties); *Galearis spectabilis*, showy orchis, and *Liparis liliifolia*, lily-leaved twayblade. This late spring array may be followed in midsummer by *Corallorhiza maculata* var. *maculata*, the spotted coralroot and perhaps, in a few selected beech woodlands, *Triphora trianthophora*, the three birds orchis. In many of these same areas autumn brings the elusive *Corallorhiza odontorhiza* var. *odontorhiza*, the autumn coralroot.

In the open copses and damp roadsides *Cypripedium parviflorum* var. *pubescens*, the large yellow lady's-slipper, may be more frequent than in the woodlands, especially in the uplands, and watch carefully in the damp ditches for *Gymnadeniopsis clavellata*, the little club-spur orchis, a common orchid eastward but very rare here. Look for the delightful *Pogonia ophioglossoides*, the rose pogonia, as well. Summer may bring a riot of color with *Platanthera ciliaris*, the orange fringed orchis, and *P. peramoena*, the purple fringeless orchis. Both of these species may grow in open woodlands as well as in sunny roadside ditches and under power lines.

Summer and autumn present a variety of ladies'-tresses including *Spiranthes cernua*, the nodding ladies'-tresses, in wet to moist open grasslands. In wooded situations the delicate *S. ovalis*, the oval ladies'-tresses, grow. Both varieties may be found in Arkansas but *S. ovalis* var. *erostellata*, the northern oval ladies'-tresses, is the more frequent, with var. *ovalis*, the southern oval ladies'-tresses, considerably rarer.

Searching the grasslands and meadows as well as old cemeteries may yield several species also found in the open prairies. Some that might be found include *Platanthera lacera*, the green fringed orchis; *Spiranthes lacera* var. *gracilis*, the southern slender ladies'-tresses; *S. tuberosa*, the little ladies'-tresses, and *S. vernalis*, the grass-leaved ladies'-tresses.

Acidic deciduous woodlands are often suitable homes for *Corallorhiza wisteriana*, Wister's coralroot; *Goodyera pubescens*, downy rattlesnake orchis; *Hexalectris spicata*, the crested coralroot; *Isotria verticillata*, the large whorled pogonia; *Malaxis unifolia*, the green adder's-mouth; *Tipularia discolor*, the crane-fly orchis, and, in southern Arkansas, the recently described *Spiranthes sylvatica*, the woodland ladies'-tresses.

Not all of the aforementioned species will occur throughout the region, and many are relatively rare. Searching with a watchful eye and an open mind may reveal some of them.

7. The Eastern Prairie Element

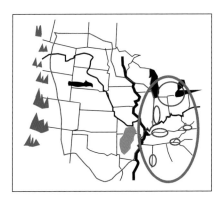

Some of the more prominent prairie elements found east of the Mississippi River include areas in eastern Mississippi/western Alabama, southern Indiana/northern Kentucky, western Ohio, southern Michigan/southwestern Ontario, northwestern Georgia, and extreme southwestern Virginia. These prairie islands are usually typified by the presence of several prairie species of grasses, legumes, composites, and, for our purposes, one or more of the short list of prairie orchids:

Calopogon oklahomensis, Oklahoma grass-pink
Cypripedium candidum, small white lady's-slipper
Platanthera leucophaea, eastern prairie fringed orchis
Spiranthes magnicamporum, Great Plains ladies'-tresses.

In most of these islands only one of the above species is present but several other orchid species that also enjoy these open, usually calcareous, mesic grasslands and bluffs are also found.
Species that may be found in one or more of the eastern prairie element areas include:

Calopogon oklahomensis, Oklahoma grass-pink
Calopogon tuberosus, common grass-pink
Corallorhiza wisteriana, Wister's coralroot
Cypripedium acaule, pink lady's-slipper
Cypripedium candidum, small white lady's-slipper
Cypripedium parviflorum var. *makasin*, northern small yellow lady's-slipper

Cypripedium parviflorum var. *pubescens,* large yellow lady's-slipper
Gymnadeniopsis clavellata, little club-spur orchis
Habenaria repens, water-spider orchis
Liparis loeselii, Loesel's twayblade
Listera australis, southern twayblade
Malaxis unifolia, green adder's-mouth
Platanthera blephariglottis, northern white fringed orchis
Platanthera ciliaris, orange fringed orchis
Platanthera cristata, orange crested orchis
Platanthera flava var. *flava,* southern tubercled orchis
Platanthera flava var. *herbiola,* northern tubercled orchis
Platanthera huronensis, green bog orchis
Platanthera lacera, green fringed orchis
Platanthera leucophaea, eastern prairie fringed orchis
Platanthera peramoena, purple fringeless orchis
Platanthera psycodes, small purple fringed orchis
Pogonia ophioglossoides, rose pogonia
Spiranthes cernua, nodding ladies'-tresses
Spiranthes lacera var. *gracilis,* southern slender ladies'-tresses
Spiranthes laciniata, lace-lipped ladies'-tresses
Spiranthes magnicamporum, Great Plains ladies'-tresses
Spiranthes odorata, fragrant ladies'-tresses
Spiranthes tuberosa, little ladies'-tresses
Spiranthes vernalis, grass-leaved ladies'-tresses

Two very small prairie islands in northwestern Georgia and extreme southwestern Virginia are both home to plants of *Spiranthes magnicamporum*, the Great Plains ladies'-tresses, although non-orchid prairie species are what first drew attention to these areas. Other orchid species found there include *Malaxis unifolia*, the green adder's-mouth, *Platanthera lacera*, the green fringed orchis, *Spiranthes cernua*, nodding ladies'-tresses, and *S. lacera* var. *gracilis*, southern slender ladies'-tresses.

A significant prairie island in eastern Mississippi and western Alabama supports large colonies of *Spiranthes magnicamporum*, the Great Plains ladies'-tresses, and in one place in Alabama *Cypripedium candidum*, the small white lady's-slipper. Additional orchids found in this prairie region include *Malaxis unifolia*, the green adder's-mouth, *Platanthera lacera*, the green fringed orchis, *Spiranthes cernua*, nodding ladies'-tresses, and *S. lacera* var. *gracilis*, the southern slender ladies'-tresses. *Calopogon oklahomensis*, the Oklahoma grass-pink, has also been documented for local sites in this region.

Northward into southern Indiana/northern Kentucky, western Ohio, and southern Michigan/southwestern Ontario the prairies are much more localized but tend to be seasonally wetter and support many additional species. *Cypripedium candidum*, the small white lady's-slipper, and *Platanthera leucophaea*, the eastern prairie fringed orchis, are found in several of these areas and both *Cypripedium parviflorum* var. *makasin*, the northern small yellow lady's-slipper, and *C. parviflorum* var. *pubescens*, the large yellow lady's-slipper, are often present as well as their hybrid with *C. candidum*—*C.* ×*andrewsii*. *Calopogon oklahomensis*, the Oklahoma grass-pink, has been documented from southern Indiana. Additional orchid species that may be found in these areas, especially in Ohio and Michigan, include *Calopogon tuberosus*, the common grass-pink; *Cypripedium acaule*, the pink lady's-slipper; *Gymnadeniopsis clavellata*, the little club-spur orchis; *Liparis loeselii*, Loesel's twayblade; *Malaxis unifolia*, the green adder's-mouth; *Platanthera blephariglottis*, the northern white fringed orchis; *P. ciliaris*, the orange fringed orchis; *P. flava* var. *herbiola*, the northern tubercled orchis; *P. lacera*, the green fringed orchis; *P. psycodes*, the small purple fringed orchis; *Pogonia ophioglossoides*, the rose pogonia; *Spiranthes cernua*, the nodding ladies'-tresses, and *S. lacera* var. *gracilis*, the southern slender ladies'-tresses. *Spiranthes magnicamporum*, the Great Plains ladies'-tresses, is rare and local in only a few areas, particularly in southern Ohio and southern Indiana.

At the Limit

Although treated to some degree in several other places in this book, there are a number of species that reach the geographic extreme of their distribution in the prairies and Great Plains region of North America. Because this region is centrally located and not bound on either the east or the west by oceans, these species in question are either primarily disjuncts or do not extend their range in any of the four geographic directions. There are an equal number of more northerly species that reach the southern limit of their range here in the Midwest. In addition, several species of a more southerly distribution reach their natural northern limit in the southern portions of the Midwest.

Those that represent southern extensions and disjunct populations, although some are found further south within the Rocky Mountains, include:

Calypso bulbosa var. *americana*, eastern fairy-slipper
Coeloglossum viride var. *virescens*, long-bracted green orchis
Corallorhiza maculata var. *occidentalis*, western spotted coralroot
Cypripedium parviflorum var. *makasin*, northern small yellow lady's-slipper
Cypripedium reginae, showy lady's-slipper
Goodyera oblongifolia, giant rattlesnake orchis
Goodyera repens, lesser rattlesnake orchis
Listera convallarioides, broad-lipped twayblade
Piperia unalascensis, Alaskan piperia
Platanthera aquilonis, northern green bog orchis
Platanthera orbiculata, pad-leaved orchis
Spiranthes casei, Case's ladies'-tresses
Spiranthes lacera var. *lacera*, northern slender ladies'-tresses
Spiranthes romanzoffiana, hooded ladies'-tresses

Those species that are primarily southern or southeastern in distribution and drift northward in the more temperate areas include:

Corallorhiza wisteriana, Wister's coralroot
Cypripedium kentuckiense, ivory-lipped lady's-slipper
Cypripedium parviflorum var. *parviflorum*, southern small yellow lady's-slipper
Gymnadeniopsis nivea, snowy orchis
Habenaria repens, water-spider orchis
Hexalectris spicata, crested coralroot
Listera australis, southern twayblade

Platanthera cristata, orange crested orchis
Platanthera flava var. *flava*, southern tubercled orchis
Platanthera peramoena, purple fringeless orchis
Spiranthes laciniata, lace-lipped ladies'-tresses
Spiranthes odorata, fragrant ladies'-tresses
Spiranthes ovalis var. *ovalis*, southern oval ladies'-tresses
Spiranthes sylvatica, woodland ladies'-tresses
Spiranthes tuberosa, little ladies'-tresses
Spiranthes vernalis, grass-leaved ladies'-tresses

Many eastern species reach the western limit of their ranges in the prairies and Great Plains region. Some of those would certainly include many of the southern and southeastern species listed above as well as:

Corallorhiza odontorhiza var. *pringlei*, Pringle's autumn coralroot
Cypripedium acaule, pink lady's-slipper
Galearis spectabilis, showy orchis
Goodyera pubescens, downy rattlesnake orchis
Gymnadeniopsis clavellata, little club-spur orchis
Isotria medeoloides, small whorled pogonia
Isotria verticillata, large whorled pogonia
Liparis liliifolia, lily-leaved twayblade
Listera australis, southern twayblade
Malaxis unifolia, green adder's-mouth
Platanthera blephariglottis, northern white fringed orchis
Platanthera ciliaris, orange fringed orchis
Platanthera flava var. *herbiola*, northern tubercled orchis
Platanthera hookeri, Hooker's orchis
Platanthera lacera, green fringed orchis
Platanthera leucophaea, eastern prairie fringed orchis
Platanthera peramoena, purple fringeless orchis
Platanthera psycodes, small purple fringed orchis
Pogonia ophioglossoides, rose pogonia
Spiranthes cernua, nodding ladies'-tresses
Spiranthes lacera var. *gracilis*, southern slender ladies'-tresses
Spiranthes lucida, shining ladies'-tresses
Spiranthes ovalis var. *erostellata*, northern oval ladies'-tresses
Spiranthes tuberosa, little ladies'-tresses
Spiranthes vernalis, grass-leaved ladies'-tresses
Tipularia discolor, crane-fly orchis
Triphora trianthophora, three birds orchis

Only a few western species reach their eastern limits in the prairies and Great Plains and deserve to be eagerly sought out in more suitable areas:

Epipactis gigantea, stream orchid
Platanthera dilatata var. *albiflora*, bog candles
Platanthera stricta, slender green bog orchis
Spiranthes diluvialis, Ute ladies'-tresses

Although their primary distribution is western, both *Corallorhiza striata* var. *vreelandii*, Vreeland's striped coralroot, and *Piperia unalascensis*, the Alaskan piperia, are found in scattered locations eastward.

Tips and Trips

Much has previously been written in this work about places to go and times to visit. Here are some tips to make your visit more pleasurable.

Orchids, as tough as they may seem at times, are vulnerable to disturbance, so tread as lightly as you are able and respect both the plants and the habitat. Throughout Canada and in much of the United States most orchids are protected, especially if they are in provincial, national, or state parks (as is all vegetation)—do not be tempted to pick or dig the plants!

Trips in late April and early May usually are best for the spring species in much of the area; late June and early July are best for the Black Hills. The summer months in the north from July to early August are best for southern Manitoba and the like. Many of the areas are popular with tourists and require advance reservations to ensure accommodations. For those traveling from the United States into Canada, for some reason unknown to most, Canadian service stations tend to stay open late into the evening but in turn open later in the morning. ALWAYS gas up the night before to avoid delays the next morning! If you are photographing, bring twice as much film as you plan to use or ample memory cards for your digital camera. Purchasing the film you want may not be a possibility while on the road. The weather is bound to be variable but a hat and sunscreen are almost always in order. In the wetlands insects may abound, so come prepared. Enjoy the regional cuisines and customs and meet wonderful local people, who are often more than happy to either grant permission to cross their lands or guide you to a special spot. Stop and talk with the local populace. They will appreciate it.

What Next?

Discovery of new species for any given area or the relocation of a "lost" species is always a great thrill. Many opportunities still present themselves for such discoveries in nearly all of the provinces and states within the prairies and Great Plains region. One need not be an orchid professional to chance upon any of these species, or the travelling orchid enthusiast may set out to specifically hunt them down.

Some of those species considered extirpated or not seen in recent years in specific states or provinces include:

Cypripedium candidum, small white lady's-slipper: Saskatchewan
Isotria medeoloides, small whorled pogonia: Missouri
Calopogon oklahomensis, Oklahoma grass-pink: Minnesota, Wisconsin, Illinois
Platanthera leucophaea, eastern prairie fringed orchis: Louisiana, Oklahoma
Platanthera praeclara, western prairie fringed orchis: Wyoming
Platanthera psycodes, small purple fringed orchis: Missouri and the curious Nebraska record
Gymnadeniopsis nivea, snowy orchis, and *Spiranthes magnicamporum*, Great Plains ladies'-tresses: Arkansas
Platanthera dilatata var. *albiflora*, bog candles: Black Hills of South Dakota
Platanthera stricta, slender bog orchis: Black Hills of South Dakota

Information on additional sites for three of the most significant prairie/plains species is always welcome: *Cypripedium candidum*, the small white lady's-slipper, *Platanthera leucophaea*, the eastern prairie fringed orchis, and *P. praeclara*, the western prairie fringed orchis.

A wish-list for new sites in new states where there is suitable habitat includes such species as *Cypripedium candidum*, small white lady's-slipper, and *Isotria medeoloides*, small whorled pogonia, in Arkansas and *Spiranthes diluvialis*, Ute ladies'-tresses, in South Dakota.

Additional locations for regional rarities more frequently seen west of the prairies and Great Plains:

Corallorhiza striata var. *vreelandii*, Vreeland's striped coralroot
Epipactis gigantea, stream orchid
Goodyera oblongifolia, giant rattlesnake orchis
Listera convallarioides, broad-lipped twayblade
Piperia unalascensis, Alaskan piperia
Platanthera dilatata var. *albiflora*, bog candles

Platanthera stricta, slender bog orchis: Black Hills of South Dakota
Spiranthes diluvialis, Ute ladies'-tresses

Range extensions for some of the southern species such as *Calopogon barbatus*, bearded grass-pink, *Pteroglossaspis ecristata*, crestless plume orchis, and *Gymnadeniopsis integra*, yellow fringeless orchis, from southern and western Louisiana, may find their way to the Cajun prairies in the northwestern part of the state.

Spiranthes praecox presents a particularly interesting problem. All the specimens that have been seen from northwestern Louisiana, Arkansas, and Oklahoma that were labeled *S. praecox*, the giant ladies'-tresses, have proven to be *S. sylvatica*, the woodland ladies'-tresses. The same holds true for some specimens from eastern Texas, although *S. praecox* is certainly present along the Gulf region of Texas as well as in southern Louisiana. *Spiranthes praecox* should be carefully sought in the prairie areas of northwestern Louisiana, Arkansas, and Oklahoma and compared with specimens of *S. sylvatica*.

All of these situations offer challenges to both the amateur and the serious native orchidophile. If you do find any of these species please remember that you should not pick any parts of the plants, but take many photos and communicate with the appropriate agency in the province or state in which you have found the plants. In this age of digital photography and electronic transmissions it becomes very easy to report such exciting findings!

Appendix 1. Distribution of the Orchids of the Prairies and Great Plains Region.

	Sas.	Man.	Ark.	Col.	Ill.	Ia.	Kan.	La.	Minn.	Mo.	Mt.	Neb.	N.D.	N.M.	Ok.	S.D.	Tex.	Wi.	Wy.
Aplectrum hyemale Adam and Eve			X		X	X			X	X					X			X	
Calopogon oklahomensis Oklahoma grass-pink			X		X	X	X	X	X	X					X		X	X	
Calopogon tuberosus common grass-pink			X		X	X		X	X	X					X		X	X	
Calypso bulbosa eastern fairy-slipper																X			
Coeloglossum viride var. *virescens* long-bracted green orchis		X			X	X			X	X		X	X					X	
Corallorhiza maculata var. *maculata* spotted coralroot	X	X			X	X			X									X	
Corallorhiza maculata var. *occidentalis* western spotted coralroot	X	X		X					X			X	X	X		X		X	
Corallorhiza odontorhiza var. *odontorhiza* autumn coralroot		X	X	X	X	X	X	X	X	X	X		X	X	X	X	X		
Corallorhiza odontorhiza var. *pringlei* Pringle's autumn coralroot						X												X	
Corallorhiza striata var. *striata* striped coralroot	X	X										X	X			X			
Corallorhiza striata var. *vreelandii* Vreeland's striped coralroot																X			X
Corallorhiza trifida early coralroot	X	X		X	X				X		X		X	X				X	
Corallorhiza wisteriana Wister's coralroot			X	X	X		X	X		X		X			X	X	X		
Cypripedium acaule pink lady's-slipper		X			X				X									X	
Cypripedium candidum small white lady's-slipper	X	X			X	X			X	X		X	X			X		X	
Cypripedium kentuckiense ivory-lipped lady's-slipper			X					X							X		X		
Cypripedium parviflorum var. *parviflorum* southern small yellow lady's-slipper		X			X		X			X		X			X				
Cypripedium parviflorum var. *makasin* northern small yellow lady's-slipper	X	X			X	X			X		X		X					X	
Cypripedium parviflorum var. *pubescens* large yellow lady's-slipper	X	X	X	X	X	X	X		X	X			X	X		X		X	X
Cypripedium reginae showy lady's-slipper	X	X	X		X	X			X	X			X					X	
Epipactis gigantea stream orchid														X	X	X	X		
Epipactis helleborine broad-leaved helleborine			X		X				X	X								X	
Galearis spectabilis showy orchis			X		X	X	X		X	X		X			X			X	

	Sas.	Man.	Ark.	Col.	Ill.	Ia.	Kan.	La.	Minn.	Mo.	Mt.	Neb.	N.D.	N.M.	Ok.	S.D.	Tex.	Wi.	Wy.
Goodyera oblongifolia giant rattlesnake orchis				X									X			X			X
Goodyera pubescens downy rattlesnake orchis			X		X	X		X	X						X			X	
Goodyera repens lesser rattlesnake orchis															X				X
Gymnadeniopsis clavellata little club-spur orchis			X		X	X	X	X		X					X		X	X	
Gymnadeniopsis nivea snowy orchis			X					X									X		
Habenaria repens water-spider orchis			X					X							X		X		
Hexalectris spicata crested coralroot			X		X		X	X		X				X	X		X		
Isotria medeoloides small whorled pogonia					X					X									
Isotria verticillata large whorled pogonia			X		X			X		X					X		X		
Liparis liliifolia lily-leaved twayblade			X		X	X			X	X					X			X	
Liparis loeselii Loesel's twayblade	X	X	X		X	X	X		X	X	X							X	X
Listera australis southern twayblade			X		X			X							X		X		X
Listera convallarioides broad-lipped twayblade																X			X
Malaxis brachypoda white adder's-mouth					X														
Malaxis unifolia green adder's-mouth		X	X		X	X	X	X		X					X		X	X	
Piperia unalascensis Alaskan piperia														X		X			
Platanthera aquilonis northern green bog orchis	X	X			X	X					X	X	X	X		X		X	X
Platanthera blephariglottis northern white fringed orchis					X														
Platanthera ciliaris orange fringed orchis			X		X			X		X					X		X		
Platanthera cristata orange crested orchis			X					X									X		
Platanthera dilatata var. *dilatata* tall white northern bog orchis		X			X						X	X				X		X	X
Platanthera dilatata var. *albiflora* bog candles																X			
Platanthera flava var. *flava* southern tubercled orchis			X		X			X		X					X		X		
Platanthera flava var. *herbiola* northern tubercled orchis			X		X	X			X	X							R	X	
Platanthera hookeri Hooker's orchis		X			X	X			X									X	
Platanthera huronensis green bog orchis		X							X							X		X	

	Sas.	Man.	Ark.	Col.	Ill.	Ia.	Kan.	La.	Minn.	Mo.	Mt.	Neb.	N.D.	N.M.	Ok.	S.D.	Tex.	Wi.	Wy.
Platanthera lacera green fringed orchis		X	X		X		X		X	X					X		X	X	
Platanthera leucophaea eastern prairie fringed orchis			X		X	X			X			X			X			X	
Platanthera orbiculata pad-leaved orchis		X			X		X				X				X				X
Platanthera peramoena purple fringeless orchis			X		X					X									
Platanthera praeclara western prairie fringed orchis		X			X	X		X	X	X		X	X		X	X			X
Platanthera psycodes small purple fringed orchis		X			X	X			X			R						X	
Platanthera stricta slender green bog orchis																X			
Pogonia ophioglossoides rose pogonia		X	X		X			X	X	X			X		X		X	X	
Spiranthes casei Case's ladies'-tresses																		X	
Spiranthes cernua nodding ladies'-tresses			X		X	X	X	X	X	X		X			X	X	X	X	
Spiranthes diluvialis Ute ladies'-tresses				X								X							X
Spiranthes lacera var. *lacera* northern slender ladies'-tresses		X	X		X	X	X		X	X								X	
Spiranthes lacera var. *gracilis* southern slender ladies'-tresses			X		X	X	X	X	X	X					X		X	X	
Spiranthes laciniata lace-lipped ladies'-tresses			0					X											
Spiranthes lucida shining ladies'-tresses					X	X	X		X	X		X							
Spiranthes magnicamporum Great Plains ladies'-tresses		X	?	X	X	X	X	X	X	X		X	X	X	X	X	X	X	
Spiranthes odorata fragrant ladies'-tresses			X					X							X		X		
Spiranthes ovalis var. *ovalis* southern oval ladies'-tresses			X					X							X		X		
Spiranthes ovalis var. *erostellata* northern oval ladies'-tresses			X		X	X	X	X		X					X		X		
Spiranthes romanzoffiana hooded ladies'-tresses	X	X			X	X			X		X	X	X	X		X		X	
Spiranthes sylvatica woodland ladies'-tresses			X					X							X		X		
Spiranthes tuberosa little ladies'-tresses			X		X		X	X		X					X		X		
Spiranthes vernalis grass-leaved ladies'-tresses			X		X	X	X	X		X		X		0	X	X	X		
Tipularia discolor crane-fly orchis			X		X			X		X					X		X		
Triphora trianthophora three birds orchid			X		X	X	X	X		X		X			X			X	X

X = state or provincial record; ? = questionable record; R = report; 0 = erroneous report

Appendix 2. Flowering Times of the Orchids of the Prairies and Great Plains Region of North America.

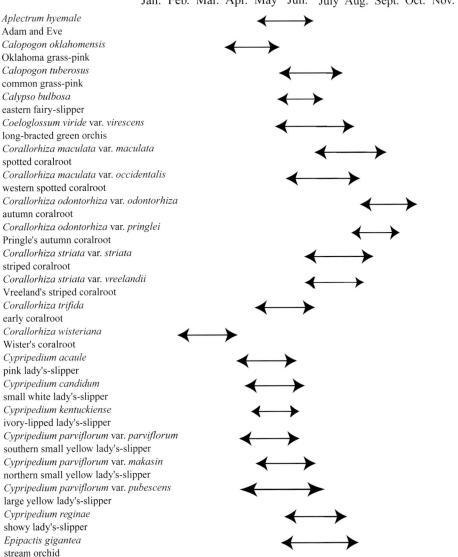

Species	Flowering Time
Epipactis helleborine broad-leaved helleborine	July–Aug.
Galearis spectabilis showy orchis	Apr.–May
Goodyera oblongifolia giant rattlesnake orchis	July–Aug.
Goodyera pubescens downy rattlesnake orchis	July–Aug.
Goodyera repens lesser rattlesnake orchis	July–Aug.
Gymnadeniopsis clavellata little club-spur orchis	June–Sept.
Gymnadeniopsis nivea snowy orchis	June–July
Habenaria repens water-spider orchis	Mar.–Nov.
Hexalectris spicata crested coralroot	June–July
Isotria medeoloides small whorled pogonia	May–June
Isotria verticillata large whorled pogonia	May–June
Liparis liliifolia lily-leaved twayblade	May–June
Liparis loeselii Loesel's twayblade	June–July
Listera australis southern twayblade	Apr.–May
Listera convallarioides broad-lipped twayblade	June–July
Malaxis brachypoda white adder's-mouth	June–July
Malaxis unifolia green adder's-mouth	May–July
Piperia unalascensis Alaskan piperia	June–July
Platanthera aquilonis northern green bog orchis	June–Aug.
Platanthera blephariglottis northern white fringed orchis	June–July
Platanthera ciliaris orange fringed orchis	June–July
Platanthera cristata orange crested orchis	June–July

Flowering Time Chart

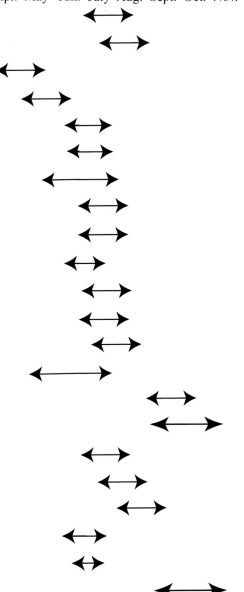

Flowering Time Chart

	Jan.	Feb.	Mar.	Apr.	May	Jun.	July	Aug.	Sept.	Oct.	Nov.	Dec.
Spiranthes odorata fragrant ladies'-tresses	←——→									←———————→		
Spiranthes ovalis var. *ovalis* southern oval ladies'-tresses									←——→			
Spiranthes ovalis var. *erostellata* northern oval ladies'-tresses									←→			
Spiranthes romanzoffiana hooded ladies'-tresses								←——→				
Spiranthes sylvatica woodland ladies'-tresses				←——→								
Spiranthes tuberosa little ladies'-tresses						←—————————→						
Spiranthes vernalis grass-leaved ladies'-tresses					←———→							
Tipularia discolor crane-fly orchis						←——→						
Triphora trianthophora three birds orchid						←——→						

Flowering Time Chart

Glossary

adventive: non-native
anterior: front or upper
anthesis: time of flowering
apomictic: fertilized within the embryo without pollination; an asexual means of reproduction
appressed: placed tightly against; opposite of divergent
approximate: (flowers) lying close by, but not overlapping
auricle: ear-like appendage
autogamous: one of several means of self-pollination
axillary: on the side
basal rosette: a cluster of leaves all arising at the base of the plant
bract: a modified leaf
calcareous: derived from limestone, limey
calciphile: a lime lover
callus: a thickened area, usually at the base of the lip
capitate: like a head; "with capitate hairs" refers to hairs with ball-like tips
carnivorous: insect eating
cauline: on the stem
chasmogamous: with fully opened, usually sexual, flowers
chlorophyll: a green pigment manufactured by the plant and essential for photosynthesis
ciliate: with short, slender hairs
circumneutral: a pH of about 6.5
clavate: club-shaped
cleistogamous: with closed flowers that are usually self-pollinating
column: the structure in an orchid that has both the anthers and the pistil
conduplicate: folded lengthwise
coralloid: coral-like
cordate: heart-shaped
coriaceous: leathery
corymb: a determinate inflorescence with all of the branches the same length and the outer flowers opening first
crenulate: with a short, wavy margin
crest: a series of ridges or a group of hairs; usually yellow or a color contrasting with the lip
cyme: a determinate inflorescence with the central flowers opening first

dentate: toothed
determinate: with a specific ending, not continuing to grow indefinitely
dilated: broadened
disjunct: occurring apart from the normal range of the species
distal: away from the center or main body; the underside
divergent: spreading or widely separated
dorsal sepal: the sepal opposite the lip; usually uppermost in most orchids
emarginate: with a short projection at the tip
endemic: native to a specific area
ephemeral: appearing for a brief time, often a few weeks each season
epiphyte, epiphytic: living in the air
erose: with an irregular margin
extant: still to be found
extirpated: no longer found
extralimital: occurring on the very edge of the species' normal range
falcate: sickle-shaped
filiform: slender and thread-like
foliaceous: leaf-like in appearance, as in foliaceous bracts
forma: a taxonomic rank that indicates a recognizable variation from the norm usually in either color or form of growth, i.e., white-flowered forms
glabrous: smooth
glaucous: with a whitish cast
habit: form or mode of growth; e.g., upright, or sprawling
habitat: the particular environment of a plant
hemi-epiphyte: usually growing on the base of trees or on logs
hybrid: the result of a sexual crossing of two different parents; indicated with a multiplication sign (×) either before the species name or between the names of the two parents
hydric: perennially wet habitat
isthmus: a narrowed portion, often at the base of the lip
keel: a ridge
lacerate: slashed
lateral sepals: the sepals positioned on the side of the flower
lax: (flowers) loosely arranged
lip: the modified third petal of an orchid
lithophytic: growing on rocks
madder: dusky-purple color
marcescent: withering but not falling off
marl: a calcareous or limey wetland
mentum: a short, rounded, thickened projection formed by the base of the petals, similar to a spur but not tapered to a point
mesic: moisture condition intermediate between wet and dry
monotypic (genus): a genus having only one species
mucronate: short, sharp point

mycotrophic: obtaining food through mycorrhizal fungi, which are attached to roots
naturalized: a non-native species reproducing in its adopted habitat
Neotropical: new world tropics
nominate, nominate variety: the pure species or variety, exclusive of subspecies, variety, or form
non-resupinate: with only a single twist so the lip is at the top
oblanceolate: narrowly oblong
obovate: broadly oblong
orbicular: rounded
orifice: opening
ovary: the female structure that produces the seed
panicle: a branching inflorescence similar to a raceme, with stalked flowers
pedicellate flowers: those flowers held on pedicels or stalks
pedicellate ovary: with ovary and flower stalk merged into one structure, typical of orchids
peduncle: stalk of a flower
perianth: the combined petals and sepals of the flower
petiole: stem portion of the leaf
plicate: soft and with many longitudinal ribs, often folded
pollinia: the structure in orchids that contains the pollen
posterior: lower or rear
pseudobulb: a swollen storage organ prominent in many epiphytic orchids and occasionally in a few genera of terrestrial orchids
puberulent: with a fine dusting of very short, soft hairs
pubescent, pubescence: downy with short, soft hairs
putative: assumed, but not scientifically proven, in reference to hybrid parentage
raceme: unbranched, indeterminate inflorescence with stalked flowers; branched racemes are technically panicles
rank: arrangement of flowers in lines or vertical rows, especially in *Spiranthes*
reflexed: bent backward
reniform: kidney-shaped
resupinate: twisted around so that the lip is lowermost
rhizome: an elongated basal stem; typically underground in terrestrials and along the strata (tree trunks, branches, rocks, etc.) in epiphytes
rhombic: with parallel sides but then tapered on both ends
rostellum: the part of the column, usually beak-shaped, that contains the stigmatic surface and to which the pollen adheres
saccate: sack-shaped
saprophytic, saprophyte: living off decaying vegetable matter
scape: a leafless stem that arises from the base of the plant
secund: all to one side
segregate genus (species): a genus or species that has been separated from another genus or species

senesces: withers
sepals: outer floral envelope
serpentine: rock formations high in toxic metals and minerals such as nickel, asbestos, and magnesium
sessile: without a stem or stalk
spatulate: oblong with a narrowed base
sphagnous: of or pertaining to sphagnum moss
spike: unbranched inflorescence with sessile or unstalked flowers
spiranthoid: a *Spiranthes* or member of a genus closely allied to *Spiranthes*
spur: a slender, tubular or sac-like structure usually formed at the base of the lip and often containing nectar
striate, striations: with stripes
subtend: beneath the flower at the point of attachment to the stem
sympatric: growing together in the same habitat
taxon, *pl.* taxa: particular taxonomic classification; i.e., subspecies, variety, species, form
tepal: sepals and petals of similar morphology
terete: rounded
terminal: at the end
transverse: growing horizontally (as opposed to parallel) to the axis (or stem)
tubercle: a thickened projection
ultramafic, ultramific: soils and substrates found in serpentine areas with very high minerals such as nickel, magnesium, and asbestos
umbel: an inflorescence with pedicels arising from the end of a stem like the spokes of an umbrella
undulate: wavy
waif: a random individual occurrence
whorl: arranged in a circle around the same point on the stem
xeric: perennially dry habitat

Bibliography

Actor, G. F. 1984. Natural hybridization of *Cypripedium candidum* and *C. calceolus*. M.S. thesis, University of North Dakota, Grand Forks.

Ames, D., P. B. Acheson, L. Heshka, B. Joyce, J. Neufeld, R. Reeves, E. Reimer, and I. Ward. 2005. *Orchids of Manitoba*. Winnipeg: Native Orchid Conservation.

Ames, O. 1910. *Orchidaceae*. Vol. 4, *The Genus Habenaria in North America*. North Easton, Mass.: Ames Botanical Laboratory.

———. 1924. *An Enumeration of the Orchids of the United States and Canada*. Boston: American Orchid Society.

Armstrong, D., M. Fritz, P. Miller, and O. Byers. 1997. Population and habitat viability assessment workshop for the western prairie fringed orchid (*Platanthera praeclara*), final report. Apple Valley, Minn.: Conservation Breeding Specialist Group.

Bateman, R. M., A. Pridgeon, and M. W. Chase. 1997. Phylogenetics of subtribe Orchidinae based on nuclear ITS sequences: 2. Intergeneric relationships and taxonomic revision to achieve monophyly of *Orchis* sensu stricto. *Lindleyana* 12: 113–41.

Bender, J. 1995. Element stewardship abstract for *Platanthera leucophaea* and *Platanthera praeclara*. Minneapolis: The Nature Conservancy.

Bentley, S. 1999. *Native Orchids of the Southern Appalachian Mountains*. Chapel Hill: University of North Carolina Press.

Bingham, M. T. 1939. *Orchids of Michigan*. Bloomfield Hills, Mich.: Cranbrook Institute of Science.

Bjugstad-Porter, R. 1993. The western prairie fringed orchid (*Platanthera praeclara*): its response to burning and associated mycorrhizal fungi. M.S. thesis, University of Wyoming, Laramie.

Borkowsky, C. 1999. Western prairie fringed orchid tour, Tolstoi, Manitoba. *North American Native Orchid Journal*. 5(1): 34–39.

Bowles, M. L. 1983. The tallgrass prairie orchids, *Platanthera leucophaea* (Nutt.) Lind. and *Cypripedium candidum* Muhl. *ex* Wild.: some aspects of their status, biology, and ecology, and implications toward management. *Natural Areas Journal* 3,4: 14–37.

Bowles, M. L., and A. Duxbury. 1986. Report on the status of *Platanthera praeclara* Sheviak & Bowles in Oklahoma, Kansas, Nebraska, South Dakota, and North Dakota. Unpublished. Denver: U.S. Fish & Wildlife Service, p. 42.

Bray, T. E., and B. L. Wilson. 1992. Status of *Platanthera praeclara* Sheviak & Bowles (western prairie fringed orchid) in the Platte River Valley in Nebraska from Hamilton to Garden counties. *Transactions of the Nebraska Academy of Sciences* 29: 57–61.

Brown, P. M. 1993. *A Field and Study Guide to the Orchids of New England and New York*. Jamaica Plain, Mass.: Orchis Press.

———. 2002. Resurrection of the genus *Gymnadeniopsis* Rydberg. *North American Native Orchid Journal* 8: 32–40.

———. 2005. Additions and emendations to *The Wild Orchids of North America, North of Mexico*. *Sida* 21(4): 2297–319.

Brown, P. M., and S. N. Folsom. 1997. *Wild Orchids of the Northeastern United States*. Ithaca: Cornell University Press.

———. 2002. *Wild Orchids of Florida*. Gainesville: University Press of Florida.

———. 2003. *The Wild Orchids of North America, North of Mexico*. Gainesville: University Press of Florida.

———. 2004. *Wild Orchids of the Southeastern United States*. Gainesville: University Press of Florida.

———. 2005. *Wild Orchids of Florida, with References to the Atlantic and Gulf Coastal Plains*, updated and expanded edition. Gainesville: University Press of Florida.

———. 2005. *Wild Orchids of the Canadian Maritimes and Northern Great Lakes Region*. Gainesville: University Press of Florida.

———. 2005. *Wild Orchids of the Pacific Northwest and Canadian Rockies*. Gainesville: University Press of Florida.

Case, F. W. 1987. *Orchids of the Western Great Lakes Region*. Revised edition, Bulletin 48. Bloomfield Hills, Mich.: Cranbrook Institute of Science.

Catling, P. M. 1989. Biology of North American representatives of the subfamily Spiranthoideae. In *North American Native Terrestrial Orchid Propagation and Production*. Chadds Ford, Penn.: Brandywine Conservancy.

Catling, P. M., and J. E. Cruise. 1974. *Spiranthes casei*, a new species from northeastern North America. *Rhodora* 76(808): 526–36.

Catling, P. M., and V. R. Brownell. 1987. New and significant vascular plant records for Manitoba. *Canadian Field Naturalist* 101(3): 437–39.

Center for Plant Conservation plant profile 9293. *Platanthera praeclara* http://www.centerforplantconservation.org/ASP/CPC_ViewProfile.asp?CPCNum=9293.

Central Iowa Orchid Society. 2005. Our Native Iowa Orchids. *http://www.c-we.com/cios/native.htm*.

Coleman, R. A. 2002. *The Wild Orchids of Arizona and New Mexico*. Ithaca: Cornell University Press.

———. 1995. *Wild Orchids of California*. Ithaca: Cornell University Press.

Collicutt, D. R. 1993. Status report on the western prairie white fringed orchid, *Platanthera praeclara*, in Canada. Ottawa: Committee on the Status of Endangered Wildlife in Canada.

Correll, D. S. 1950. *Native Orchids of North America*. Waltham, Mass.: Chronica Botanica.

Cribb, P. 1997. *The Genus Cypripedium*. Portland, Ore.: Timber Press.

Cuthrell, D. L. 1994. Insects associated with the prairie fringed orchids, *Platanthera praeclara* Sheviak & Bowles and *P. leucophaea* (Nuttall) Lindley. M.S. thesis, North Dakota State University, Fargo.

Duxbury, A. J. 1988. The western prairie fringed orchid (*Platanthera praeclara* Sheviak & Bowles) in the Great Plains with particular reference to North Dakota. *Proceedings of the North Dakota Academy of Science* 42: 4.

Fitzpatrick, T. J., and M. F. L. Fitzpatrick. 1900. The Orchidaceae of Iowa. *Proceedings of the Iowa Academy of Sciences* 7: 187–96.

[FNA] Flora of North America Editorial Committee, eds. 1993. *Flora of North America, North of Mexico.* Vol. 26, *Orchidaceae.* New York: Oxford University Press.

Freeman, R. 1998. *Cypripedium* hybrids in Mahnomen County, Minnesota. *North American Native Orchid Journal.* 4(4): 333–35.

Fritz, M. 1993. Nebraska's threatened and endangered species: western prairie fringed orchid. Lincoln: Nebraska Game and Parks Commission, 6 pp.

From, M. M., and P. Read. 1998. *Platanthera praeclara* strategies for conservation and propagation. *North American Native Orchid Journal* 4(4): 299–332.

Fuller, A. 1933. Studies on the flora of Wisconsin. Part 1: The orchids; Orchidaceae. *Bulletin of the Public Museum of Milwaukee* 14(1): 1–248.

Garay, L. A. 1982. A generic revision of the Spiranthinae. *Botanical Museum Leaflet* (Harvard University) 28(4): 277–425.

Hapeman, J. R. 1996. Orchids of Wisconsin. http://www.botany.wisc.edu/Orchids/Orchids_of_Wisconsin.html

Henderson, J. 1977. *A Taxonomic Treatment of the Genus* Habenaria *in Eleven Midwestern States.* M.S. thesis, University of Missouri-Kansas City.

Heshka, L. 2003. Field identification of the *Platanthera hyperborea* complex in Manitoba. *Native Orchid News* 5(4).

———. 2005. Manitoba Orchids. http://www.Manitobaorchidsociety.ca/native/ManitobaNativeOrchids/index.html#to.

Homoya, M. A. 1993. *Orchids of Indiana.* Bloomington and Indianapolis: Indiana Academy of Science, University of Indiana Press.

Hornbeck, J. H., C. H. Sieg, D. J. Reyer, and D. J. Bacon. 2003. Conservation Assessment for the Large Round-leaved Orchid in the Black Hills National Forest, South Dakota and Wyoming. Custer, S. Dak.: U.S. Department of Agriculture, Forest Service.

International Code of Botanical Nomenclature (St. Louis Code). Prepared and edited by W. Greuter, J. McNeill, F. R. Barrie, H.-M. Burdet, V. DeMoulin, T. S. Filgueiras, D. H. Nicolson, P. C. Silva, J. E. Skog, P. Trehane, N. J. Turland, and D. L. Hawksworth. 2000. *Regnum Vegetabile* 138. http://www.bgbm.fu-berlin.de/iapt/nomenclature/code/SaintLouis/0000St.Luistitle.htm.

[IPNI] International Plant Names Index. 2004. http://www.ipni.org/index.html [accessed Nov. 1, 2005].

Irwin, F. G., and P. Reeder. 1965. Wild Orchids of Illinois. *Outdoor Illinois* 4(10): 4–14.

Johnson, G. P. 2000. An addition and a deletion to the Orchidaceae of Arkansas. *North American Native Orchid Journal* 6(2): 140–41.

———. 2004. An Annotated checklist of the orchids of Arkansas. *Claytonia* 23(4): 5–6.

Johnson, G., and C. Slaughter. n.d. *Wild Orchids of Arkansas* (in preparation).

Johnson, K. L. 1981. Rare orchids III: the rare bog orchids (Genus *Platanthera*) of Manitoba. *Bulletin of the Manitoba Naturalists' Society* 4(8): 7.

———. 1986. Yet more about orchids!!! The western prairie fringed orchid (*Platanthera praeclara*). *Bulletin of the Manitoba Naturalists' Society* 11(10): 12.

Johnson, S. 1996. Orchids of Louisiana's Cajun Prairie. *North American Native Orchid Journal* 2(4): 368–71.

Kartesz, J. T. 1994. *A Synonymized Checklist of the Vascular Plants of United States, Canada and Greenland*, 2nd ed. 2 vols. Portland, Ore.: Timber Press.

Kartesz, J. T., and C. A. Meacham, editors. 1999. *Synthesis of the North American Flora*, ver. 1.0. Chapel Hill: North Carolina Botanical Garden.

Kaul, R. B. 1986. In R. L. McGregor, T. M. Barkley, R. E. Brooks, E. K. Schofield, W. T. Barker. M. Bolick, S. P. Churchill, R. L. Hartman, R. B. Kaul, O. A. Kolstad, G. E. Larson, D. M. Sutherland, T. Van Bruggen, R. R. Weedon, and D. H. Wilken, *Flora of the Great Plains*, 1268–84. Lawrence, Kans.: University Press of Kansas.

Keenan, P. E. 1999. *Wild Orchids Across North America*. Portland, Ore.: Timber Press.

Kravig, M. L. 1969. Orchids of the Black Hills. *Proceeding of the South Dakota Academy of Sciences* 48: 119–31.

Liggio, J., and A. O. Liggio. 1999. *Wild Orchids of Texas*. Austin: University of Texas Press.

Long, J. C. 1970. *Native Orchids of Colorado*. Museum Pictorial 16. Denver: Denver Museum of Natural History.

Luer, C. A. 1975. *The Native Orchids of the United States and Canada excluding Florida*. Bronx: New York Botanical Garden.

Magrath, L. W. 1971. Native Orchids of Kansas. *Transactions of the Kansas Academy of Science* 74: 287–309.

———. 1973. *Orchids of the Prairies and Plains Regions of North America*. Ph.D. dissertation, University of Kansas, Lawrence.

———. 1989. Nomenclatural notes on *Calopogon*, *Corallorhiza*, and *Cypripedium* (Orchidaceae) in the Great Plains region. *Sida* 13: 371–72.

———. 2001. Native Orchids of Oklahoma. *Crosstimbers* (Spring): 18–25.

Magrath, L. K., and J. Taylor. 1978. Orchids and other new and interesting plants from Oklahoma. In *New, Rare, and Infrequently Collected Plants in Oklahoma*, publication 2. Durant: Herbarium of Southeastern Oklahoma State University.

Menges, E. S., and P. F. Quintana Ascencio. 1997. Preliminary analysis of plant dormancy in the western prairie fringed orchid. Informal report to Nancy P. Sather, Minnesota Department of Natural Resources.

Norby, M. M. 1997. Horticulturists hoping to increase native orchid's numbers. *Research Nebraska*. 8(2): 6–13.

O'Neil, D. C. 1995. The Buffalo Orchid. *AOS Bulletin* 64(4): 372–74.

Petrie, W. 1981. *Guide to the Orchids of North America*. Blaine, Wash.: Hancock House.

Pleasants, J. M., and K. Klier. 1995. Genetic variation within and among populations of the eastern and western prairie fringed orchids, *Platanthera leucophaea* and *P. praeclara*. Report to the Iowa Department of Natural Resources.

Pleasants, J. M., and S. Moe. 1993. Floral display size and pollination of the western prairie fringed orchid, *Platanthera praeclara* (Orchidaceae). *Lindleyana*. 8(1): 32.

Reddoch, J. M., and A. Reddoch. 1997. The orchids of the Ottawa District. *Canadian Field-Naturalist* 111(1): 1–186.

Rydberg, P. A. 1901. *Gymnadeniopsis* Rydberg. In N. L. Britton, *Manual of the Flora of the Northern States and Canada*, 293. New York: H. Holt and Co.

———. 1965. *Flora of the Prairies and Plains of Central North America*. New York: Hafner.

Sather, N. 1991. *Western Prairie Fringed Orchid: a threatened Midwestern prairie plant*. St. Paul: Minnesota Department of Natural Resources.

Sather, N., E. Menges, and P. F. Quintana Ascencio. 1997. *Platanthera praeclara* in Minnesota: summary of status and monitoring results for 1996. St. Paul: Minnesota Department of Natural Resources.

Scoggin, H. J. 1978. *The Flora of Canada*, Part 2. Ottawa: National Museum Natural Science Publications in Botany 7.

Sheviak, C. J. 1974. *An Introduction to the Ecology of the Illinois Orchidaceae*. Springfield: Illinois State Museum.

———. 1982. *Biosystematic Study of the Spiranthes cernua Complex*. Bulletin 448. Albany: New York State Museum.

———. 1990. A new form of *Cypripedium montanum* Douglas ex Lindley. *Rhodora* 92: 47–49.

———. 1991. Morphological variation in the compilospecies *Spiranthes cernua* (L.) L.C. Rich.: ecologically limited effects of gene flow. *Lindleyana* 6: 228–34.

———. 1994. *Cypripedium parviflorum* Salisbury part 1: The smaller flowered plants. *American Orchid Society Bulletin* 63(60): 664–69.

———. 1995. *Cypripedium parviflorum* Salisbury part 2: The larger flowered plants and patterns of variation. *American Orchid Society Bulletin* 64(6): 606–12.

———. 1999. The identities of *Platanthera hyperborea* and *P. huronensis*, with the description of a new species from North America. *Lindleyana* 14: 193–203.

Sheviak, C. J., and M. L. Bowles. 1986. The prairie fringed orchids: a pollinator-isolated species pair. *Rhodora*. 88: 267–90.

Sheyenne Ranger District. 1999. Management guidelines for the western prairie fringed orchid on the Sheyenne National Grassland, Dakota Prairie Grasslands. Lisbon, N. Dak.

Sieg, C. H., and R. M. King. 1995. Influence of environmental factors and preliminary demographic analyses of a threatened orchid, *Platanthera praeclara*. *American Midland Naturalist* 134(2): 307–23.

Sieg, C. H. 1997. The mysteries of a prairie orchid. *Endangered Species Bulletin* 22(4): 12–13.

Slaughter, C. R. 1993. *Wild Orchids of Arkansas*. Morrilton, Ark.: Privately published.

Smith, W. R. 1993. *Orchids of Minnesota*. Minneapolis: University of Minnesota Press.

South Dakota Orchid Society. 2005. Orchids of South Dakota. http://nathist.sdstate.edu/Orchids/nativeorchids.htm

Stoutamire, W. P. 1974. Relationships of purple fringed orchids *Platanthera psycodes* and *P. grandiflora*. *Brittonia* 26: 42–58.

Summers, W. 1987. *Missouri Orchids*, 2nd ed. Jefferson City: Missouri Department of Conservation.

Thurman, C. M.. and E. E. Hickey. 1989. A Missouri survey of six species of federal concern: auriculate false foxglove, *Tomanthera auriculata*; Mead's milkweed, *Asclepias meadii*; geocarpon, *Geocarpon minimum*; Missouri bladder-pod, *Lesquerella filiformis*; western prairie fringed orchid, *Platanthera praeclara*; and decurrent false aster, *Boltonia decurrens*. Missouri Department of Conservation.

U.S. Fish and Wildlife Service. 1994. Western prairie fringed orchid: *Platanthera praeclara*. Pierre, S.D.

———. 1996. *Platanthera praeclara* (western prairie fringed orchid) recovery plan. Ft. Snelling, Minn.

Wallace, L. E. 2002. An evaluation of taxonomic boundaries in *Platanthera dilatata* (Orchidaceae). *Rhodora* 105(924): 322–36.

———. 2003. Molecular evidence for allopolyploid speciation and recurrent origins in *Platanthera huronensis* (Orchidaceae). *International Journal of Plant Science* 164(6): 907–16.

———. 2004. A comparison of genetic variation and structure in the allopolyploid *Platanthera huronensis* and its diploid progenitors, *Platanthera aquilonis* and *Platanthera dilatata* (Orchidaceae). *Canadian Journal of Botany* 82: 244–52.

Watson, L. E. 1989. Status survey of *Platanthera praeclara*, western white fringed prairie orchid, in Oklahoma. Norman: Oklahoma Natural Heritage Inventory.

Watson, W. C. 1993. Inventory of northern Iowa for *Platanthera leucophaea* (Nuttall) Lindley and *Platanthera praeclara* Sheviak & Bowles: final report, 1992. Des Moines: U.S. Fish and Wildlife Service and the Iowa Department of Natural Resources.

Weber, S. 1996. *Spiranthes cernua* and *S. casei* on a farm in Wisconsin. *North American Native Orchid Journal* 2(1): 57–63.

Williams, J. G., A. E. Williams, and N. Arlott. 1983. *A Field Guide to Orchids of North America*. New York: Universe Books.

Winterringer, G. S. 1967. *Wild Orchids of Illinois*. Springfield: Illinois State Museum.

Witt, W. 2006. *Orchids in Your Pocket: A Guide to the Native Orchids of Iowa*. Iowa City: University of Iowa Press.

Wolken, P. M. 1995. *Habitat and life history of the western prairie fringed orchid (Platanthera praeclara)*. University of Wyoming.

Wood, J. 1986. *Calypso bulbosa* var. *occidentalis* and var. *speciosa*. *Kew Magazine* 3(4): 147–51.

Zettler, L. W., K. A. Piskin, S. L. Stewart, J. J. Hartsock, M. L. Bowles, and T. Bell. 2005. Protocorm mycobionts of a federally threatened orchid, *Platanthera leucophaea*, and a technique to promote leaf formation in seedlings. *Studies in Mycology* 53: 163–71.

Photo Credits

All photographs were taken by Paul Martin Brown, except for the following, which were generously loaned by those credited.

Larry Magrath, *Cypripedium parviflorum* var. *parviflorum* (front cover); forma *albolabium*

Jack Price, *Cypripedium kentuckiense* forma *pricei*

Bill Summers, *Cypripedium kentuckiense* forma *summersii*, and *Platanthera peramoena* forma *doddsiae*

Lorne Heshka, *Platanthera praeclara* pink form

Elaine Ebbert, *Corallorhiza* 'ochroleuca'

Index

Primary entries for taxa are in **bold**. Page numbers for color photographs are in *italics*.

Adam-and-Eve, **18**, *19*, 110, 219, 228–33, 236, 238, 243, 252, 286, 289, 297
   yellow-flowered form, **18**, *19*, 219
Adder's-mouth
   white, **120**, *121*, 230, 242–43
   green, 121, **122**, *123*, 222, 227–28, 230–33, 236, 238–39, 241, 243, 246, 249, 252, 283, 287, 293, 297, 301, 304
      two-leaved form, **122**, *123*, 222
      variegated-leaf form, **122**, *123*, 222
Adiantum capillus-veneris, 72
Amesia
   gigantea, 262, 266
   latifolia forma *monotropoides*, 74
Alaskan orchid, **126**, *127*, 223, 235, 237, 242–44, 263, 269–71, 293, 295, 303, 307
   dwarf form, **126**, 223
Andrews' hybrid fringed orchis, 152, 164, **168**, 224
Andrews' hybrid lady's-slipper, 56, 64, **68**, 221
Aplectrum, 17
   hyemale, **18**, *19*, 110, 219, 228–33, 236, 238, 243, 252, 286, 289, 297
      forma *pallidum*, **18**, *19*, 219
   spicatum var. *pallidum*, 18
Arethusa bulbosa, 214
Autumn coralroot, 34, **40**, *41*, 219, 228–33, 235–37, 249, 252–53, 277, 286, 292, 297
   yellow-stemmed form, **40**, *41*, 219

Bearded grass-pink, 22, *214*, 249, 308
Bicolor hybrid fringed orchis, 136
Big Horns (Wyoming), 295
Black Hills (South Dakota/Wyoming), 72, 118, 126, 292–94
Blephariglottis
   blephariglottis, 264, 266

   ciliaris, 264, 266
   cristata, 264, 266
   leucophaea, 265
   peramoena, 265–66
   psycodes, 266
Bog candles, 130, 140, **142**, *143*, 223, 237, 242–43, 253, 264, 268, 270, 293, 296, 307
Bog orchis
   green, 130, 131, *132*, 140, **150**, *151*, 223, 228, 233, 237, 239, 241, 243, 256, 260, 264, 268, 270, 278, 283, 290, 293, 301
   northern green, 130, **131**, *132*, *133*, 140, 150, 166, 223, 227, 230–31, 234–37, 239, 241, 243–44, 248, 256, 263, 269–70, 278, 283, 293, 295, 303
   slender, 4, 118, 129, **166**, *167*, 169, 224, 237, 242–44, 268–70, 293, 305, 307–8
   tall white northern, 130, **140**, *141*, 150, 223, 227, 230, 233–34, 237, 239, 243, 253, 264, 268, 270, 293, 295, 305
Broad-leaved helleborine, 71, **74**, *75*, 221, 228–29, 233, 262, 268
   albino form, **74**, *75*, 221
   green-flowered form, **74**, *75*, 221
   variegated form, **74**, *75*, 221
   white-flowered form, **74**, *75*, 221
   yellow-flowered form, **74**, *75*, 221
Broad-lipped twayblade, 115, **118**, *119*, 222, 237, 242–43, 253, 293, 295, 303, 307
   three-leaved form, **118**, 222

Calopogon, 21, 172
   barbatus, 22, *214*, 249, 308
   oklahomensis, 21, **22**, *23*, 214, 219, 228–33, 236–38, 241, 244, 247–49, 252, 255, 275, 283, 286, 300, 307

forma *albiflorus*, 12, **22**, *23*
pulchellus, 259, 261, 266
tuberosus
  var. *tuberosus*, 21, 22, **24**, *25*, 214, 219, 228–30, 232–33, 236–38, 241, 243, 246–49, 252, 259, 261, 267, 283, 286, 300
  forma *albiflorus*, **24**, *25*, 219
  var. *simpsonii*, 24
Calypso bulbosa, 27
  var. *americana*, 4, 27, **28**, *29*, 118, 219, 237, 242. 253. 261, 268, 292–95, 303
    forma *albiflora*, **28**, *29*, 219
    forma *biflora*, **28**, *29*, 219
    forma *rosea*, **28**, *29*, 219
  var. *bulbosa*, 27
  var. *occidentalis*, 27
  var. *speciosa*, 27
Case's ladies'-tresses, 174, **176**, *177*, 223, 239, 242, 244, 256, 270–71, 303
Channell's hybrid fringed orchis, 136, 138, **168**, 224
Chatterbox, 71, **72**, *73*, 221, 237, 244, 247, 252–53, 262, 266, 269, 289, 292, 295, 305, 307
  yellow-flowered, **72**, *73*, 221
  red-leaved, **72**, 221
Chiwaulkee Prairie (Wisconsin), 284
Cleistes, 171
Club-spur orchis, little, 89, **90**, *91*, 221, 228–29, 231–33, 236–38, 241, 248–50, 252, 256, 263, 268, 283, 286, 293, 297, 301, 304
  white-flowered form (Slaughter's), **90**, *91*, 221
  spurless form (Wright's), **90**, 221
Coeloglossum, 31, 95, 260
  bracteatum, 261, 266
  viride
    var. *interjecta*, 31
    var. *virescens*, 31, 32, 33, 219, 229–30, 233–35, 237, 249, 261, 267,-69, 289, 292, 295, 297, 303
    var. *viride*, 31
Common grass-pink, 21, 22, **24**, *25*, 214, 219, 228–30, 232–33, 236–38, 241, 243, 246–49, 252, 259, 261, 267, 283, 286, 300
  white-flowered form, **24**, *25*, 219

Cooper's cranberry coralroot, **50**, *51*, 220
Corallorhiza, 34
  bigelovii, 262, 266
  corallorhiza, 266
  maculata
    subsp. *occidentalis*, 261, 266
    var. *immaculata*, 38
    var. *intermedia*, 38
    var. *maculata*, 34, 35, **36**, *37*, 38, 219, 227, 229–30, 233, 238, 248, 261, 267, 277, 286, 289, 292, 295, 297
      forma *flavida*, **36**, *37*, 219
      forma *rubra*, **36**, *37*, 219
    var. *mexicana*, 36
    var. *occidentalis*, 34, 36, **38**, *39*, 219, 227, 229, 233–35, 237–38, 244, 255, 261, 267, 275, 277, 289, 292, 295, 303
      forma *aurea*, **38**, *39*, 219
      forma *immaculata*, **38**, *39*, 219
      forma *intermedia*, **38**, 219
      forma *punicea*, **38**, *39*, 219
    var. *punicea*, 38
  multiflora, 261, 266
    var. *flavida*, 36
  ochroleuca, 44, 46
  odontorhiza
    var. *odontorhiza*, 34, **40**, *41*, 219, 228–33, 235–37, 249, 252–53, 277, 286, 292, 297
      forma *flavida*, **40**, *41*, 219
    var. *pringlei*, 34, **42**, *43*, 219, 228–33, 242, 244, 255, 261, 267, 277, 304
  pringlei, 261, 266
  striata
    var. *striata*, 34, **44**, *45*, 220, 227, 234–35, 237, 242, 246, 278, 289, 292, 295
      forma *eburnea*, **44**, *45*, 220
      forma *fulva*, **44**, *45*, 220
    var. *vreelandii*, 34, 44, **46**, *47*, 220, 237, 239, 242, 244, 262, 267, 278, 2889, 292, 295, 307
      forma *flavida*, **46**, *47*, 220
    var. *flavida*, 46
  trifida, 34, **48**, *49*, 220, 227, 229, 233–35, 237–38, 242, 253, 267, 289, 292, 295
    forma *verna*, **48**, *49*, 220
    var. *verna*, 48

verna, 48
wisteriana, 34, **50**, *51*, 220, 228–29, 231–37, 286, 289, 297, 300, 303
    forma *albolabia*, **50**, *51*, 220
    forma *cooperi*, **50**, *51*, 220
    forma *rubra*, **50**, *51*, 220
Coralroot
    autumn, 34, **40**, *41*, 219, 228–33, 235–37, 249, 252–53, 277, 286, 292, 297
        yellow-stemmed form, **40**, *41*, 219
    crested, 99, **100**, *101*, 222, 228–29, 232–33, 235–36, 238, 247–49, 252, 286, 297, 303
        albino form (Wilder's), **100**, 222
        white-lipped form, **100**, 222
        yellow-flowered form, **100**, *101*, 222
    Pringle's autumn, 34, **42**, *43*, 219, 228–33, 242, 244, 255, 261, 267, 277, 304
    early, 34, **48**, *49*, 220, 227, 229, 233–35, 237–38, 242, 253, 267, 289, 292, 295
    spotted, 34, 35, **36**, *37*, 38, 219, 227, 229–30, 233, 238, 248, 261, 267, 277, 286, 289, 292, 295, 297
        red-stemmed form, 36, *37*, 219
        yellow-stemmed form, 36, *37*, 219
    striped, 34, **44**, *45*, 220, 227, 234–35, 237, 242, 246, 278, 289, 292, 295
        yellow/white form, **44**, *45*, 220
    Vreeland's striped, 34, 44, **46**, *47*, 220, 237, 239, 242, 244, 262, 267, 278, 2889, 292, 295, 307
        yellow/white form, **46**, *47*, 220
    western spotted, 34, 36, **38**, *39*, 219, 227, 229, 233–35, 237–38, 244, 255, 261, 267, 275, 277, 289, 292, 295, 303
        brown-stemmed form, **38**, 219
        golden yellow/spotted form, **38**, *39*, 219
        red-stemmed form, **38**, *39*, 219
        yellow spotless form, **38**, *39*, 219
    Wister's, 34, **50**, *51*, 220, 228–29, 231–37, 286, 289, 297, 300, 303
        cranberry colored form (Cooper's), **50**, *51*, 220
        red-stemmed form, **50**, *51*, 220
        yellow-stemmed/white-lipped form, **50**, *51*, 220
Crane-fly orchis, 207, **208**, *209*, 226, 229–30, 232, 234, 238, 244, 250, 252, 287, 298, 304
    green-leaved form, **208**, 226
Crested coralroot, 99, **100**, *101*, 222, 228–29, 232–33, 235–36, 238, 247–49, 252, 286, 297, 303
    albino form (Wilder's), **100**, 222
    white-lipped form, **100**, 222
    yellow-flowered form, **100**, *101*, 222
Crested orchis, orange, 129, **138**, *139*, 223, 228, 232, 238, 241, 243, 247, 267–68, 287, 301, 303
    pale-flowered form, **138**, *139*, 223
Crestless plume orchid, **215**
Cypress Hills of Alberta/Saskatchewan, 295
Cypripedium, 53
    acaule, 53, **54**, *55*, 220, 233, 238, 242–43, 248, 262, 268, 300, 304
        forma *albiflorum*, **54**, *55*, 220
        forma *biflorum*, **54**, 220
        forma *lancifolium*, **54**, *55*, 220
    calceolus, 252–53, 260, 262, 266
        var. *planipetalum*, 247, 262
    candidum, 5, 53, **56**, *57*, 60, 64, 220, 227, 229–30, 233–35, 237–38, 241, 246, 248, 250, 253, 283, 300, 307
    daultonii, 262, 266
    flavescens, 262, 266
    kentuckiense, 53, **58**, *59*, 220, 228, 236, 242–44, 247, 249, 252, 255, 262, 267, 275, 297, 303
        forma *pricei*, **58**, *59*, 220
        forma *summersii*, **58**, *59*, 220
    parviflorum, 56, 260
        var. *makasin*, 53, **62**, *63*, 220, 227, 229–30, 233–35, 238, 241–42, 248, 255, 267–68, 278, 283, 292, 300, 303
        var. *parviflorum*, 53, **60**, *61*, 64, 220, 228–29, 231, 233–34, 236, 243, 252, 255, 262, 267, 275, 278, 286, 289, 297, 303
        forma *albolabium*, **60**, *61*, 220
        var. *planipetalum*, 267
        var. *pubescens*, 53, 62, **64**, *65*, 220, 227–29, 231, 233, 235, 237–39, 241, 247, 254, 256, 262, 267–68, 278, 283, 286, 289, 292, 295, 297, 301

pubescens, 262, 268
   var. *makasin*, 268
reginae, 53, **66**, *67*, 221, 227–29, 231, 233, 235, 238, 242–43, 246–48, 250, 262, 268, 283, 303
   forma *albolabium*, **66**, *67*, 221
   spectabile, 262, 268
   veganum, 262, 268
   ×*andrewsii* nm *andrewsii*, 56, 64, **68**, 221
   ×*andrewsii* nm *favillianum*, 62, **68**, 221
   ×*andrewsii* nm *landonii*, **68**, 221
   ×*favillianum*, 68
   ×**herae**, **69**
   ×*landonii*, 68
Cytherea bulbosa, 261, 268

Dactylorhiza, 31
   viridis, 261, 268
Dodds' purple fringeless orchis, 129, **158**, *159*, 224
Downy rattlesnake orchis, 81, **84**, *85*, 221, 228–29, 231, 233, 238, 243, 286, 297, 304
Dwarf rattlesnake orchis, 81, **86**, *87*, 221, 228–29, 231, 233, 238, 242–43, 247, 292, 295, 303

Early coralroot, 34, **48**, *49*, 220, 227, 229, 233–35, 237–38, 242, 253, 267, 289, 292, 295
Eastern fairy-slipper, 4, 27, **28**, *29*, 118, 219, 237, 242, 253, 261, 268, 292–95, 303
   pink-flowered form, 28, *29*, 219
   two-flowered form, 28, *29*, 219
   white-flowered form, 28, *29*, 219
Eastern prairie fringed orchis, 5, 12, 56, 129, **154**, *155*, 169, 224, 230–32, 234–36, 239, 241, 244, 248, 250, 252–53, 256, 265, 269, 279, 283, 287, 300–301, 307
Epipactis, 71
   gigantea, 71, **72**, *73*, 221, 237, 244, 247, 252–53, 262, 266, 269, 289, 292, 295, 305, 307
      forma *citrina*, 72, *73*, 221
      forma *rubrifolia*, 72, 221
   helleborine, 71, **74**, *75*, 221, 228–29, 233, 262, 268

      forma *alba*, **74**, *75*, 221
      forma *luteola*, **74**, *75*, 221
      forma *monotropoides*, **74**, *75*, 221
      forma *variegata*, **74**, 221
      forma *viridens*, **74**, *75*, 221
   latifolia, 262, 268
      var. *alba*, 74
      var. *variegata*, 74

Fairy-slipper, eastern, 4, 27, **28**, *29*, 118, 219, 237, 242, 253, 261, 268, 292–95, 303
   pink-flowered form, 28, *29*, 219
   two-flowered form, 28, *29*, 219
   white-flowered form, 28, *29*, 219
Faville's hybrid yellow lady's-slipper, 62, **68**, 221
Fen orchid, 109, **112**, *113*, 222, 227–28, 230–31, 233, 235, 239, 241, 243, 246–47, 249–50, 253, 283, 286, 293, 301
Fissipes acaulis, 262, 268
Fragrant ladies'-tresses, 175, 178, 190, **192**, *193*, 194, 225, 229, 232, 236, 238, 242–43, 247, 264, 269–70, 287, 301, 303
Fringed orchis
   Andrews' hybrid, 152, 164, **168**, 224
   bicolor hybrid, 136
   Channell's hybrid, 136, 138, **168**, 224
   crested, 129, **138**, *139*, 223, 228, 232, 238, 241, 243, 247, 267–68, 287, 301, 303
      pale-flowered form, **138**, *139*, 223
   eastern prairie, 5, 12, 56, 129, **154**, *155*, 169, 224, 230–32, 234–36, 239, 241, 244, 248, 250, 252–53, 256, 265, 269, 279, 283, 287, 300–301, 307
   green, 129, **152**, *153*, 164, 169, 223, 228, 230, 233–34, 236, 239, 241, 244, 246, 249, 252, 265, 269, 283, 287, 293, 298, 301
   Holland River hybrid, 152, 154, 169
   northern white, 129, **134**, *135*, 136, 223, 230, 233, 241, 243, 263, 266, 268, 301, 304
      entire-lip form, **134**, 223
   orange, 129, **136**, *137*, 138, 223, 228, 230, 232–33, 236, 238, 241, 248, 250, 252, 264, 267–68, 287, 290, 297, 301

Reznicek's hybrid, 154, 162, 169
small purple, 129, 152, 161, **162**, *163*, *164*, 169, 224, 228, 230–31, 233–34, 239, 241, 244, 246, 248, 250, 256, 267, 269, 283, 301, 304, 307
  entire-lip form, 162, *164*, 224
  pink-flowered form, 162, *164*, 224
  spurless form, 162, 224
  white-flowered form, 162, *164*, 224
western prairie, 129, 152, 161, **162**, *163*, *164*, 169, 224, 228, 230–31, 233–34, 239, 241, 244, 246, 248, 250, 256, 267, 269, 283, 301, 304, 307
Fringeless orchis, purple, 5, 129, **158**, *159*, 224, 228, 230, 234, 247, 256, 265, 267, 269, 282, 298, 301, 303–4
  white-flowered form (Dodds'), 129, **158**, *159*, 224

Galearis, 77
  spectabilis, **78**, *79*, 110, 221, 228–29, 231, 233–34, 238, 268, 270, 286, 289, 297, 304
    forma *gordinierii*, **78**, *79*, 221
    forma *willeyi*, **78**, *79*, 221
Galeorchis spectabilis, 263, 268
  forma *gordinierii*, 78
Giant ladies'-tresses, 200, 247, 252, 266, 308
Giant rattlesnake orchis, 81, **82**, *83*, 221, 229, 236–37, 239, 242–43, 268, 290, 292, 295, 303, 307
  reticulated-leaved form, **82**, *83*, 221
Goodyera, 81
  decipiens, 263, 268
  oblongifolia, 81, **82**, *83*, 221, 229, 236–37, 239, 242–43, 268, 290, 292, 295, 303, 307
    forma *reticulata*, **82**, *83*, 221
    var. *reticulata*, 82
  pubescens, 81, **84**, *85*, 221, 228–29, 231, 233, 238, 243, 286, 297, 304
  repens, 81, **86**, *87*, 221, 228–29, 231, 233, 238, 242–43, 247, 292, 295, 303
    forma *ophioides*, 81, **86**, *87*, 221
    var. *ophioides*, 86
Gordinier's showy orchis, **78**, *79*, 221

Grass-leaved ladies'-tresses, 174, 176, **204**, *205*, 225, 228, 230–32, 234–36, 238, 242, 244, 248–49, 253, 269–71, 287, 290, 298, 301, 303–4
Grass-pink, bearded, 22, *214*, 249, 308
  common, 21, 22, **24**, *25*, 214, 219, 228–30, 232–33, 236–38, 241, 243, 246–49, 252, 259, 261, 267, 283, 286, 300
  white-flowered form, **24**, *25*, 219
  Oklahoma, 21, **22**, *23*, 214, 219, 228–33, 236–38, 241, 244, 247–49, 252, 255, 275, 283, 286, 300, 307
  white-flowered form, 12, **22**, *23*
Great Plains ladies'-tresses, 5, 175, 178, 180, **190**, *191*, 225, 228, 230–39, 241, 244, 246–47, 250, 253, 256, 279, 287, 290, 293, 296, 300–301
Green adder's-mouth, 121, **122**, *123*, 222, 227–28, 230–33, 236, 238–39, 241, 243, 246, 249, 252, 283, 287, 293, 297, 301, 304
  two-leaved form, **122**, *123*, 222
  variegated-leaf form, **122**, *123*, 222
green bog orchis, 130, 131, *132*, 140, **150**, *151*, 223, 228, 233, 237, 239, 241, 243, 256, 260, 264, 268, 270, 278, 283, 290, 293, 301
  northern, 130, **131**, *132*, *133*, 140, 150, 166, 223, 227, 230–31, 234–37, 239, 241, 243–44, 248, 256, 263, 269–70, 278, 283, 293, 295, 303
Green fringed orchis, 129, **152**, *153*, 164, 169, 223, 228, 230, 233–34, 236, 239, 241, 244, 246, 249, 252, 265, 269, 283, 287, 293, 298, 301
Ground orchid, low, 215
Gymnadeniopsis, 89, 95, 260
  clavellata var. ***clavellata***, 89, **90**, *91*, 221, 228–29, 231–33, 236–38, 241, 248–50, 252, 256, 263, 268, 283, 286, 293, 297, 301, 304
    forma *slaughteri*, **90**, *91*, 221
    forma *wrightii*, **90**, 221
    var. *ophioglossoides*, 90
  integra, *214*, 256
  nivea, 89, **92**, *93*, 22, 228, 232, 237, 241, 243, 247, 256, 263, 269–70, 303, 307
Gyrostachys stricta, 263, 268

Habenaria, 89, 92, 125, 128, 259, 261
   blephariglottis, 263, 268
   borealis var. *albiflora*, 264, 268
      var. *viridiflora*, 265, 268
   bracteata, 261, 268
   ciliaris, 264, 268
   clavellata, 263, 268
      var. *ophioglossoides*, 263, 268
      var. *wrightii*, 90, 263, 268
   cristata, 264, 268
   dilatata
      var. *dilatata*, 264, 268
      var. *albiflora*, 264, 268
   flava, 264, 268
      var. *herbiola*, 264, 268
      var. *virescens*, 264, 269
   gracilis, 265
   herbiola, 268
   hookeri, 269
      var. *abbreviata*, 148
   huronensis, 264
   hyperborea, 166, 263–64
      var. *huronensis*, 264, 269
   lacera, 265, 269
   leucophaea, 265, 269
      var. *praeclara*, 265, 269
   nivea, 263, 269
   orbiculata, 265, 269
      var. *lehorsii*, 156
      var. *longifolia*, 156
      var. *trifolia*, 156
   peramoena, 265, 269
   psycodes, 265, 269
      forma *albiflora*, 163
      var. *ecalcarata*, 163
      var. *fernaldii*, 163
      var. *grandiflora*
      var. *varians*, 163
   repens, **96**, *97*, 222, 228, 232, 236, 238, 243, 247, 252, 286, 301, 303
   saccata, 166, 265, 269
   unalascensis, 263, 269
   viridis, 269
      var. *bracteata*, 261, 269
   ×*andrewsii*
   ×*bicolor*
   ×*chapmanii*
   ×*media*, 264, 268
Helleborine gigantea, 262, 269
Helleborine, broad-leaved, 71, **74**, *75*, 221, 228–29, 233, 262, 268
   albino form, **74**, *75*, 221
   green-flowered form, **74**, *75*, 221
   variegated-leaved form, **74**, 221
   white-flowered form, **74**, *75*, 221
   yellow-flowered form, **74**, *75*, 221
Herminium (*foetida*), 126
Hexalectris, 99
   spicata, 99, **100**, *101*, 222, 228–29, 232–33, 235–36, 238, 247–49, 252, 286, 297, 303
      forma *albolabia*, 100, 222
      forma *lutea*, 100, *101*, 222
      forma *wilderi*, 100, 222
      var. *arizonica*, 100
Holland River hybrid fringed orchis
Hooded ladies'-tresses, 175, 180, 188, **198**, *199*, 225, 227–28, 230–31, 233–37, 239, 242–43, 249, 266, 268–69, 271, 279, 284, 290, 293, 296, 303
Hooker's orchis, 128, **148**, *149*, 223, 228, 230–31, 233, 239, 243, 246, 248, 268, 293
   dwarfed form, **148**, 223
   narrow-leaved form, **148**, *149*, 223
Hybrid fringed orchis
   Andrews', 152, 164, **168**, 224
   bicolor, 136
   Channell's, 136, 138, **168**, 224
   Holland River, 152, 154, 169
   Reznicek's, 154, 162, 169
Hybrid ladies'-tresses
   Ichetucknee, ***206***, 225
   intermediate**,** 192, 194, 196, ***206***, 225
   Simpson's, 198, ***206***, 225
Hybrid lady's-slipper
   Andrews', 56, 64, **68**, 221
   Faville's, 62, **68**, 221
   Landon's, **68**, 221
   Queen Hera's, **68**
Hybrid twayblade, Jones', ***114***, 222

Ibidium beckii, 266, 269
   cernuum, 265, 269

gracile, 266, 269
laciniatum, 266, 269
odoratum, 266, 269
plantagineum, 266, 269
praecox, 266, 269
stricta, 266, 269
vernale, 266, 269
Ichetucknee Springs hybrid ladies'-tresses, **206**, 225
Illinois Beach State Park, 284
Intermediate hybrid ladies'-tresses, 192, 194, 196, **206**, 225
Isotria, 103, 171
  affinis, 263, 269
  medeoloides, 103, **104**, *105*, 106, 222, 230, 233, 242–44, 263, 269, 270, 304, 307
  verticillata, 103, **106**, *107*, 222, 228, 230, 232–33, 238, 243, 248, 250, 252, 263, 270, 286, 297, 304
Ivory-lipped lady's-slipper, 53, **58**, *59*, 220, 228, 236, 242–44, 247, 249, 252, 255, 262, 267, 275, 297, 303
  concolorous yellow-flowered form (Summers'), **58**, *59*, 220
  white-flowered form (Price's), **58**, *59*, 220

Jones' hybrid twayblade, *114*, 222
Jug orchid, 215

Lace-lipped ladies'-tresses, 174, **186**, *187*, 225, 242, 266, 269, 287, 301, 303
Ladies'-tresses
  Case's, 174, **176**, *177*, 223, 239, 242, 244, 256, 270–71, 303
  fragrant, 175, 178, 190, **192**, *193*, 194, 225, 229, 232, 236, 238, 242–43, 247, 264, 269–70, 287, 301, 303
  giant, 200, 247, 252, 266, 308
  grass-leaved, 174, 176, **204**, *205*, 225, 228, 230–32, 234–36, 238, 242, 244, 248–49, 253, 269–71, 287, 290, 298, 301, 303–4
  Great Plains, 5, 175, 178, 180, **190**, *191*, 225, 228, 230–39, 241, 244, 246–47, 250, 253, 256, 279, 287, 290, 293, 296, 300–301
  hooded, 175, 180, 188, **198**, *199*, 225, 227–28, 230–31, 233–37, 239, 242–43, 249, 266,268–69, 271, 279, 284, 290, 293, 296, 303
  Ichetucknee hybrid, 192, 194, 196, **206**, 225
  intermediate hybrid, **206**, 225
  lace-lipped, 174, **186**, *187*, 225, 242, 266, 269, 287, 301, 303
  little, 175, **202**, *203*, 225, 229–30, 232, 234, 236, 238, 241, 244, 269, 287, 298, 301, 303–4
  nodding, 175, **178**, *179*, 190, 223, 238, 230–39, 241, 244, 253, 256, 265, 269–70, 284, 287, 290, 293, 298, 301, 304
    cleistogamous race, **178**, *179*
  northern oval, 175, **196**, *197*, 225, 229–32, 234, 236, 238, 244, 249, 256, 275, 287, 290, 298, 304
  northern slender, 174, **182**, *183*, 184, 198, 225, 228–31, 233–34, 239, 241, 243–44, 249, 266, 270, 280, 284, 290, 293, 303
  shining, 175, **188**, *189*, 225, 230–31, 233–35, 242, 244, 247–48, 266, 269–70, 290, 304
  Simpson's hybrid, 198, **206**, 225
  southern oval, 175, 190, 192, **194**, *195*, 196, 225, 229, 232, 236, 238, 242–44, 298, 303
  southern slender, 175, 183, **184**, *185*, 225, 229, 230–34, 236, 238–39, 241, 244, 249, 266, 269–70, 280, 284, 287, 290, 293, 298, 301, 304
  Ute, 174, **180**, *181*, 225, 235, 239, 241–42, 244, 247, 254, 256, 265, 270, 275, 279, 296, 305, 308
  woodland, 175, **200**, *201*, 225, 229, 232, 236, 238, 243–44, 247, 252, 256, 275, 287, 298, 303, 308
Lady's-slipper
  Andrews' hybrid, 56, 64, **68**, 221
  Faville's hybrid, 62, **68**, 221
  ivory-lipped, 53, **58**, *59*, 220, 228, 236, 242–44, 247, 249, 252, 255, 262, 267, 275, 297, 303
    concolorous yellow-flowered form (Summers'), **58**, *59*, 220
    white-flowered form (Price's), **58**, *59*, 220
  Landon's hybrid, **68**, 221

large yellow, 53, 62, **64**, *65*, 220, 227–29, 231, 233, 235, 237–39, 241, 247, 254, 256, 262, 267–68, 278, 283, 286, 289, 292, 295, 297, 301

northern small yellow, 53, **62**, *63*, 220, 227, 229–30, 233–35, 238, 241–42, 248, 255, 267–68, 278, 283, 292, 300, 303

pink, 53, **54**, *55*, 220, 233, 238, 242–43, 248, 262, 268, 300, 304

narrow-leaved form, **54**, *55*, 220

two-flowered form, **54**, 220

white-flowered form, **54**, *55*, 220

Queen Hera's hybrid, **68**

showy lady's-slipper, 53, **66**, *67*, 221, 227–29, 231, 233, 235, 238, 242–43, 246–48, 250, 262, 268, 283, 303

white-flowered form, **66**, *67*, 221

small white, 5, 53, **56**, *57*, 60, 64, 220, 227, 229–30, 233–35, 237–38, 241, 246, 248, 250, 253, 283, 300, 307

southern small yellow, 53, **60**, *61*, 64, 220, 228–29, 231, 233–34, 236, 243, 252, 255, 262, 267, 275, 278, 286, 289, 297, 303

white-lipped form, **60**, *61*, 220

Landon's hybrid lady's-slipper, **68, 221**

Large whorled pogonia, 103, **106**, *107*, 222, 228, 230, 232–33, 238, 243, 248, 250, 252, 263, 270, 286, 297, 304

Large yellow lady's-slipper, 53, 62, **64**, *65*, 220, 227–29, 231, 233, 235, 237–39, 241, 247, 254, 256, 262, 267–68, 278, 283, 286, 289, 292, 295, 297, 301

Lesser rattlesnake orchis, 81, **86**, *87*, 221, 228–29, 231, 233, 238, 242–43, 247, 292, 295, 303

white-veined leaf form, 81, **86**, *87*, 221

Lily-leaved twayblade, 109, **110**, *111*, 222, 228, 230–31, 233, 236, 239, 243, 252, 286, 297, 304

green-flowered form, **110**, *111*, 222

Limnorchis

dilatata, 270

media, 264

stricta, 269

Limodorum

pulchellum, 259, 261, 270

tuberosum, 259, 261, 270

Liparis, 109

liliifolia, 109, **110**, *111*, 222, 228, 230–31, 233, 236, 239, 243, 252, 286, 297, 304

forma ***viridiflora***, **110**, *111*, 222

loeselii, 109, **112**, *113*, 222, 227–28, 230–31, 233, 235, 239, 241, 243, 246–47, 249–50, 253, 283, 286, 293, 301

×***jonesii***, **114**, 222

Listera, 115

australis, 115, **116**, *117*, 222, 228, 230, 232, 236, 238, 241, 243, 252, 263, 270, 286, 301, 303, 304

forma ***scottii***, **116**, *117*, 222

forma ***trifolia***, **116**, 222

forma ***viridis***, **116**, *117*, 222

convallarioides, 115, **118**, *119*, 222, 237, 242–43, 253, 293, 295, 303, 307

forma ***trifolia***, **118**, 222

Little club-spur orchis, 89, **90**, *91*, 221, 228–29, 231–33, 236–38, 241, 248–50, 252, 256, 263, 268, 283, 286, 293, 297, 301, 304

white-flowered form (Slaughter's), **90**, *91*, 221

spurless form (Wright's), **90**, 221

Little ladies'-tresses, 175, **202**, *203*, 225, 229–30, 232, 234, 236, 238, 241, 244, 269, 287, 298, 301, 303–4

Loesel's twayblade, 109, **112**, *113*, 222, 227–28, 230–31, 233, 235, 239, 241, 243, 246–47, 249–50, 253, 283, 286, 293, 301

Long-bracted green orchis, 31, 32, 33, 219, 229–30, 233–35, 237, 249, 261, 267–69, 289, 292, 295, 297, 303

Low ground orchid, 215

Lysias orbiculata var. *pauciflora*, 156

Malaxis, 109, 120

brachypoda, **120**, *121*, 230, 242, 243

unifolia, 120, **122**, *123*, 222, 227–28, 230–33, 236, 238–39, 241, 243, 246, 249, 252, 283, 287, 293, 297, 301, 304

forma ***bifolia***, **122**, *123*, 222

forma ***variegata***, **122**, *123*, 222

Microstylis unifolia, 270

Moccasin flower, 53, **54**, *55*, 220, 233, 238,

242–43, 248, 262, 268, 300, 304
narrow-leaved form, **54**, *55*, 220
two-flowered form, **54**, 220
white-flowered form, **54**, *55*, 220

Neottia, 115
  australis, 263, 270
  convallarioides, 263, 270
  gracilis, 263, 270
  lacera, 263, 270
Nodding ladies'-tresses, 175, **178**, *179*, 190, 223, 238, 230–39, 241, 244, 253, 256, 265, 269–70, 284, 287, 290, 293, 298, 301, 304
  cleistogamous race, **178**, *179*
Nodding pogonia, 211, **212**, *213*, 226, 229–32, 234–35, 237–39, 244, 249–50, 252, 287, 298, 304
Northern bog orchis, tall white, 130, **140**, *141*, 150, 223, 227, 230, 233–34, 237, 239, 243, 253, 264, 268, 270, 293, 295, 305
Northern green bog orchis, 130, **131**, *132*, *133*, 140, 150, 166, 223, 227, 230–31, 234–37, 239, 241, 243–44, 248, 256, 263, 269–70, 278, 283, 293, 295, 303
Northern oval ladies'-tresses, 175, **196**, *197*, 225, 229–32, 234, 236, 238, 244, 249, 256, 275, 287, 290, 298, 304
Northern slender ladies'-tresses, 174, **182**, *183*, 184, 198, 225, 228–31, 233–34, 239, 241, 243–44, 249, 266, 270, 280, 284, 290, 293, 303
Northern small yellow lady's-slipper, 53, **62**, *63*, 220, 227, 229–30, 233–35, 238, 241–42, 248, 255, 267–68, 278, 283, 292, 300, 303
Northern tubercled orchis, 12, 128, **146**, *147*, 223, 228, 230, 231, 233–34, 238, 241, 243, 250, 253, 268–69, 283, 287, 298, 301
  yellow-flowered form, **146**, *147*, 223
Northern white fringed orchis, 129, **134**, *135*, 136, 223, 230, 233, 241, 243, 263, 266, 268, 301, 304
  entire-lip form, **134**, 223

Oklahoma grass-pink, 21, **22**, *23*, 214, 219, 228–33, 236–38, 241, 244, 247–49, 252, 255, 275, 283, 286, 300, 307
white-flowered form, 12, **22**, *23*
Orange crested orchis, 129, **138**, *139*, 223, 228, 232, 238, 241, 243, 247, 267–68, 287, 301, 303
  pale yellow-flowered form, **138**, *139*, 223
Orange fringed orchis, 129, **136**, *137*, 138, 223, 228, 230, 232–33, 236, 238, 241, 248, 250, 252, 264, 267–68, 287, 290, 297, 301
Orchis leucophaea, 12
  spectabilis, 259, 270
  forma *willeyi*, 78
Oval ladies'-tresses
  northern, 175, **196**, *197*, 225, 229–32, 234, 236, 238, 244, 249, 256, 275, 287, 290, 298, 304
  southern, 175, 190, 192, **194**, *195*, 196, 225, 229, 232, 236, 238, 242–44, 298, 303

Pad-leaved orchis, 129, **156**, *157*, 243, 246, 253–54, 265, 269, 293, 303
  dwarfed form, 156, 243
  few-flowered form, 156, 243
  three-leaved form, 156, 243
Physurus querceticola
Pink lady's-slipper, 53, **54**, *55*, 220, 233, 238, 242–43, 248, 262, 268, 300, 304
  narrow-leaved form, **54**, *55*, 220
  two-flowered form, **54**, 220
  white-flowered form, **54**, *55*, 220
Piperia, 125, 260
  dilatata, 264, 270
    var. *albiflora*, 270
  unalascensis, **126**, *127*, 223, 235, 237, 242–44, 263, 269–71, 293, 295, 303, 307
    forma *olympica*, **126**, *127*, 223
Platanthera, 89, 92, 95, 125–26, 128, 261
  aquilonis, 130, **131**, *132*, *133*, 140, 150, 166, 223, 227, 230–31, 234–37, 239, 241, 243–44, 248, 256, 263, 269–70, 278, 283, 293, 295, 303
    forma *alba*, **132**, 223
  blephariglottis, 129, **134**, *135*, 136, 223, 230, 233, 241, 243, 263, 266, 268, 301, 304
    forma *holopetala*, **134**, 223
  ciliaris, 129, **136**, *137*, 138, 223, 228, 230,

232–33, 236, 238, 241, 248, 250, 252, 264, 267–68, 287, 290, 297, 301
clavellata, **90**, *91*, 250, 263, 270
   forma *slaughteri*, **90**, *91*
conspicua, 136
cristata, 129, **138**, *139*, 223, 228, 232, 238, 241, 243, 247, 267–68, 287, 301, 303
   forma *straminea*, **138**, *139*, 223
dilatata
   var. *dilatata*, 130, **140**, *141*, 150, 223, 227, 230, 233–34, 237, 239, 243, 253, 264, 268, 270, 293, 295, 305
   var. *albiflora*, 130, 140, **142**, *143*, 223, 237, 242–43, 253, 264, 268, 270, 293, 296, 307
   var. *chlorantha*, 270
   var. *leucostachys*, 140
flava
   var. *flava*, 128, **144**, *145*, 146, 223, 228, 230, 232, 234, 241, 243, 247–48, 250, 252, 264, 268, 287, 298, 301, 303
   var. *herbiola*, 12, 128, **146**, *147*, 223, 228, 230, 231, 233–34, 238, 241, 243, 250, 253, 268–69, 283, 287, 298, 301
      forma *lutea*, **146**, *147*, 223
foetida, 126, 263, 270
holopetala, 134
hookeri, 128, **148**, *149*, 223, 228, 230–31, 233, 239, 243, 246, 248, 268, 293
   forma *abbreviata*, **148**, 223
   forma *oblongifolia*, **148**, *149*, 223
   var. *abbreviata*, 148
   var. *oblongifolia*, 148
huronensis, 130, 131, *132*, 140, **150**, *151*, 223, 228, 233, 237, 239, 241, 243, 256, 260, 264, 268, 270, 278, 283, 290, 293, 301
hyperborea, 131, 260, 293
   var. *alba*, 131
   var. *huronensis*, 270
lacera, 129, **152**, *153*, 164, 169, 223, 228, 230, 233–34, 236, 239, 241, 244, 246, 249, 252, 265, 269, 283, 287, 293, 298, 301
   var. *terrae-novae*, 152
leucophaea, 5, 12, 56, 129, **154**, *155*, 169,

224, 230–32, 234–36, 239, 241, 244, 248, 250, 252–53, 256, 265, 269, 279, 283, 287, 300–301, 307
nivea, 263, 270
orbiculata, 129, **156**, *157*, 243, 246, 253–54, 265, 269, 293, 303
   forma *lehorsii*, 156
   forma *longifolia*, 156
   forma *pauciflora*, 156
   forma *trifolia*, 156
   var. *lehorsii*, 156
   var. *longifolia*, 156
   var. *trifolia*, 156
macrophylla, 148, 156–57
peramoena, 5, 129, **158**, *159*, 224, 228, 230, 234, 247, 256, 265, 267, 269, 282, 298, 301, 303–4
   forma *doddsiae*, 129, **158**, *159*
praeclara, 129, 150, 154, **160**, *161*, 162, 224, 228, 231, 233–37, 239, 241, 244, 246, 248, 250, 252–53, 256, 275, 279, 283, 287, 296, 307
   pink-flowered form, 161, *162*
psycodes, 129, 152, 161, **162**, *163*, *164*, 169, 224, 228, 230–31, 233–34, 239, 241, 244, 246, 248, 250, 256, 267, 269, 283, 301, 304, 307
   forma *albiflora*, 162, *164*, 224
   forma *ecalcarata*, 162, 224
   forma *fernaldii*, 162, 224
   forma *rosea*, 162, *164*, 224
   forma *varians*, 162, 224
repens, 270
saccata, 270
stricta, 4, 118, 129, **166**, *167*, 169, 224, 237, 242–44, 268–70, 293, 305, 307–8
unalascensis, 263, 270
×*andrewsii*, 152, 164, **168**, 224
×*bicolor*, 136
×*canbyi*
×*channellii*, 136, 138, **168**, 224
×*correllii*, 166, 169
×*estesii*, 166, 169
×*hollandiae*, 152, 154, 169
×*lueri*, 136
×*media*, 140, 150, 264, 270

×*reznicekei*, 154, 162, 169
Platythelys querceticola, 215
Plume orchid, crestless, *215*, 308
Ponthieva racemosa, 214
Pogonia, 171
   affinis, 263, 270
   ophioglossoides, 171, **172**, *173*, 224, 228, 230, 232–34, 236, 238–39, 244, 246–47, 250, 252, 284, 287, 298, 301, 304
      forma **albiflora**, **172**, *173*, 224
      forma **brachypogon**, **172**, *173*, 224
      var. *brachypogon*, 173
   verticillata, 263, 270
Pogonia
   large whorled, 103, **106**, *107*, 222, 228, 230, 232–33, 238, 243, 248, 250, 252, 263, 270, 286, 297, 304
   nodding, 211, **212**, *213*, 226, 229–32, 234–35, 237–39, 244, 249–50, 252, 287, 298, 304
   rose, 171, **172**, *173*, 224, 228, 230, 232–34, 236, 238–39, 244, 246–47, 250, 252, 284, 287, 298, 301, 304
      white-flowered form, **172**, *173*, 224
      short-bearded form, **172**, *173*, 224
   small whorled, 103, **104**, *105*, 106, 222, 230, 233, 242–44, 263, 269, 270, 304, 307
Prairie fringed orchis
   eastern, 5, 12, 56, 129, **154**, *155*, 169, 224, 230–32, 234–36, 239, 241, 244, 248, 250, 252–53, 256, 265, 269, 279, 283, 287, 300–301, 307
   western, 5, 129, **158**, *159*, 224, 228, 230, 234, 247, 256, 265, 267, 269, 282, 298, 301, 303–4
Price's ivory-lipped lady's-slipper, **58**, *59*, 220
Pringle's autumn coralroot, 34, **42**, *43*, 219, 228–33, 242, 244, 255, 261, 267, 277, 304
Pseudorchis, 95
Pteroglossaspis ecristata, *215*, 308
Purple fringed orchis, small, 129, 152, 161, **162**, *163*, *164*, 169, 224, 228, 230–31, 233–34, 239, 241, 244, 246, 248, 250, 256, 267, 269, 283, 301, 304, 307
   entire-lip form, 162, *164*, 224

pink-flowered form, 162, *164*, 224
spurless form, 162, 224
white-flowered form, 162, *164*, 224
Purple fringeless orchis, 5, 129, **158**, *159*, 224, 228, 230, 234, 247, 256, 265, 267, 269, 282, 298, 301, 303–4
   white-flowered form, 129, **158**, *159*, 224
Puttyroot, **18**, *19*, 110, 219, 228–33, 236, 238, 243, 252, 286, 289, 297
   yellow-flowered form, **18**, *19*, 219

Queen Hera's hybrid lady's-slipper, **68**

Ragged orchis, 129, **152**, *153*, 164, 169, 223, 228, 230, 233–34, 236, 239, 241, 244, 246, 249, 252, 265, 269, 283, 287, 293, 298, 301
Rattlesnake orchis
   downy, 81, **84**, *85*, 221, 228–29, 231, 233, 238, 243, 286, 297, 304
   dwarf (lesser), 81, **86**, *87*, 221, 228–29, 231, 233, 238, 242–43, 247, 292, 295, 303
   giant, 81, **82**, *83*, 221, 229, 236–37, 239, 242–43, 268, 290, 292, 295, 303, 307
      reticulated form, **82**, *83*, 221
   Menzies', 81, **82**, *83*, 221, 229, 236–37, 239, 242–43, 268, 290, 292, 295, 303, 307
Rose pogonia, 171, **172**, *173*, 224, 228, 230, 232–34, 236, 238–39, 244, 246–47, 250, 252, 284, 287, 298, 301, 304
   short-bearded form, **172**, *173*, 224
   white-flowered form, **172**, *173*, 224

Scott's southern twayblade, **116**, *117*, 222
Shining ladies'-tresses, 175, **188**, *189*, 225, 230–31, 233–35, 242, 244, 247–48, 266, 269–70, 290, 304
Showy lady's-slipper, 53, **66**, *67*, 221, 227–29, 231, 233, 235, 238, 242–43, 246–48, 250, 262, 268, 283, 303
   white-flowered form, **66**, *67*, 221
Showy orchis, **78**, *79*, 110, 221, 228–29, 231, 233–34, 238, 268, 270, 286, 289, 297, 304
   pink-flowered form (Willey's), **78**, *79*, 221
   white-flowered form (Gordinier's), **78**, *79*, 221
Simpson's hybrid Ladies'-tresses, 198, ***206***, 225

Slaughter's little club-spur orchis, **90**, *91*, 221
Slender bog orchis, 4, 118, 129, **166**, *167*, 169, 224, 237, 242–44, 268–70, 293, 305, 307–8
Slender ladies'-tresses
   southern, 175, 183, **184**, *185*, 225, 229, 230–34, 236, 238–39, 241, 244, 249, 266, 269–70, 280, 284, 287, 290, 293, 298, 301, 304
   northern, 174, **182**, *183*, 184, 198, 225, 228–31, 233–34, 239, 241, 243–44, 249, 266, 270, 280, 284, 290, 293, 303
Small purple fringed orchis, 129, 152, 161, **162**, *163*, 164, 169, 224, 228, 230–31, 233–34, 239, 241, 244, 246, 248, 250, 256, 267, 269, 283, 301, 304, 307
   entire-lip form, 162, *164*, 224
   pink-flowered form, 162, *164*, 224
   spurless form, 162, 224
   white-flowered form, 162, *164*, 224
Small white lady's-slipper, 5, 53, **56**, *57*, 60, 64, 220, 227, 229–30, 233–35, 237–38, 241, 246, 248, 250, 253, 283, 300, 307
Small whorled pogonia, 103, **104**, *105*, 106, 222, 230, 233, 242–44, 263, 269, 270, 304, 307
Small yellow lady's-slipper
   northern, 53, **62**, *63*, 220, 227, 229–30, 233–35, 238, 241–42, 248, 255, 267–68, 278, 283, 292, 300, 303
   southern, 53, **60**, *61*, 64, 220, 228–29, 231, 233–34, 236, 243, 252, 255, 262, 267, 275, 278, 286, 289, 297, 303
   white-lipped form, **60**, *61*, 220
Snowy orchis, 89, **92**, *93*, 22, 228, 232, 237, 241, 243, 247, 256, 263, 269–70, 303, 307
Southern oval ladies'-tresses, 175, 190, 192, **194**, *195*, 196, 225, 229, 232, 236, 238, 242–44, 298, 303
Southern slender ladies'-tresses, 175, 183, **184**, *185*, 225, 229, 230–34, 236, 238–39, 241, 244, 249, 266, 269–70, 280, 284, 287, 290, 293, 298, 301, 304
Southern small yellow lady's-slipper, 53, **60**, *61*, 64, 220, 228–29, 231, 233–34, 236, 243, 252, 255, 262, 267, 275, 278, 286, 289, 297, 303
   white-lipped form, **60**, *61*, 220
Southern tubercled orchis, 128, **144**, *145*, 146, 223, 228, 230, 232, 234, 241, 243, 247–48, 250, 252, 264, 268, 287, 298, 301, 303
Southern twayblade, 115, **116**, *117*, 222, 228, 230, 232, 236, 238, 241, 243, 252, 263, 270, 286, 301, 303, 304
   green-flowered form, **116**, *117*, 222
   many-leaved form (Scott's), **116**, *117*, 222
   three-leaved form, **116**, 222
Spiranthes, 174
   beckii, 202, 266, 270
   casei, 174, **176**, *177*, 223, 239, 242, 244, 256, 270–71, 303
   cernua, 175, **178**, *179*, 190, 223, 238, 230–39, 241, 244, 253, 256, 265, 269–70, 284, 287, 290, 293, 298, 301, 304
      cleistogamous race, **178**, *179*
   diluvialis, 174, **180**, *181*, 225, 235, 239, 241–42, 244, 247, 254, 256, 265, 270, 275, 279, 296, 305, 308
   gracilis, 264, 270
   grayi, 202, 266, 271
   intermedia, 271
   lacera
      var. *gracilis*, 175, 183, **184**, *185*, 225, 229, 230–34, 236, 238–39, 241, 244, 249, 266, 269–70, 280, 284, 287, 290, 293, 298, 301, 304
      var. *lacera*, 174, **182**, *183*, 184, 198, 225, 228–31, 233–34, 239, 241, 243–44, 249, 266, 270, 280, 284, 290, 293, 303
   laciniata, 174, **186**, *187*, 225, 242, 266, 269, 287, 301, 303
   lucida, 175, **188**, *189*, 225, 230–31, 233–35, 242, 244, 247–48, 266, 269–70, 290, 304
   magnicamporum, 5, 175, 178, 180, **190**, *191*, 225, 228, 230–39, 241, 244, 246–47, 250, 253, 256, 279, 287, 290, 293, 296, 300–301
   ochroleuca, 178, 190
   odorata, 175, 178, 190, **192**, *193*, 194, 225, 229, 232, 236, 238, 242–43, 247, 264, 269–70, 287, 301, 303

ovalis
- var. *ovalis*, 175, 190, 192, **194**, *195*, 196, 225, 229, 232, 236, 238, 242–44, 298, 303
- var. *erostellata*, 175, **196**, *197*, 225, 229–32, 234, 236, 238, 244, 249, 256, 275, 287, 290, 298, 304
- *plantaginea*, 269, 271
- *praecox*, 200, 247, 252, 266, 308
- *romanzoffiana*, 175, 180, 188, **198**, *199*, 225, 227–28, 230–31, 233–37, 239, 242–43, 249, 266, 268–69, 271, 279, 284, 290, 293, 296, 303
  - var. *diluvialis*, 265, 271
- *simplex*, 266
- *sinensis*, 174
- *stricta*, 271
- *sylvatica*, 175, **200**, *201*, 225, 229, 232, 236, 238, 243–44, 247, 252, 256, 275, 287, 298, 303, 308
- *tuberosa*, 175, **202**, *203*, 225, 229–30, 232, 234, 236, 238, 241, 244, 269, 287, 298, 301, 303–4
  - var. *grayi*, 271
- *vernalis*, 174, 176, **204**, *205*, 225, 228, 230–32, 234–36, 238, 242, 244, 248–49, 253, 269–71, 287, 290, 298, 301, 303–4
- ×*intermedia*, **206**, 225
- ×*itchetuckneensis*, 192, 194, 196, **206**, 225
- ×*simpsonii*, 198, **206**, 225

Spotted coralroot, 34, 35, **36**, *37*, 38, 219, 227, 229–30, 233, 238, 248, 261, 267, 277, 286, 289, 292, 295, 297
- red-stemmed form, 36, *37*, 219
- yellow-stemmed form, 36, *37*, 219
- western, 34, 36, **38**, *39*, 219, 227, 229, 233–35, 237–38, 244, 255, 261, 267, 275, 277, 289, 292, 295, 303
  - brown-stemmed form, **38**, 219
  - golden yellow/spotted form, **38**, *39*, 219
  - red-stemmed form, **38**, *39*, 219
  - yellow spotless form, **38**, *39*, 219

Stream orchid, 71, **72**, *73*, 221, 237, 244, 247, 252–53, 262, 266, 269, 289, 292, 295, 305, 307
- red-leaved form, **72**, 221

yellow-flowered form, **72**, *73*, 221

Striped coralroot, 34, **44**, *45*, 220, 227, 234–35, 237, 242, 246, 278, 289, 292, 295
- yellow/white form, **44**, *45*, 220
- Vreeland's, 34, 44, **46**, *47*, 220, 237, 239, 242, 244, 262, 267, 278, 289, 292, 295, 307
  - yellow/white form, **46**, *47*, 220

Summers' ivory-lipped lady's-slipper, **58**, *59*, 220

Three birds orchid, 211, **212**, *213*, 226, 229–32, 234–35, 237–39, 244, 249–50, 252, 287, 298, 304
- blue-flowered form, **212**, 226
- multicolored form, **212**, 226
- white-flowered form, **212**, *213*, 226

Tipularia, 207
- *discolor*, 207, **208**, *209*, 226, 229–30, 232, 234, 238, 244, 250, 252, 287, 298, 304
  - forma *viridifolia*, **208**, 226

Tolstoi, Manitoba, 283

Triphora, 171, 211
- *trianthophora*, 211, **212**, *213*, 226, 229–32, 234–35, 237–39, 244, 249–50, 252, 287, 298, 304
  - forma *albidoflava*, **212**, *213*, 226
  - forma *caerulea*, **212**, 226
  - forma *rossii*, **212**, 226
  - subsp. *mexicana*, 212

Tubercled orchis
- northern, 12, 128, **146**, *147*, 223, 228, 230, 231, 233–34, 238, 241, 243, 250, 253, 268–69, 283, 287, 298, 301
  - yellow-flowered form, **146**, *147*, 223
- southern, 128, **144**, *145*, 146, 223, 228, 230, 232, 234, 241, 243, 247–48, 250, 252, 264, 268, 287, 298, 301, 303

Twayblade
- Jones' hybrid, **114**, 222
- lily-leaved, 109, **110**, *111*, 222, 228, 230–31, 233, 236, 239, 243, 252, 286, 297, 304
  - green-flowered form, **110**, *111*, 222
- Loesel's, 109, **112**, *113*, 222, 227–28, 230–31, 233, 235, 239, 241, 243, 246–47, 249–50, 253, 283, 286, 293, 301

Scott's southern, **116**, *117*, 222
southern, 115, **116**, *117*, 222, 228, 230, 232, 236, 238, 241, 243, 252, 263, 270, 286, 301, 303, 304
    green-flowered form, **116**, *117*, 222
    many-leaved form (Scott's), **116**, *117*, 222
    three-leaved form, **116**, 222

Urn orchid, 215
Ute ladies'-tresses, 174, **180**, *181*, 225, 235, 239, 241–42, 244, 247, 254, 256, 265, 270, 275, 279, 296, 305, 308

Vreeland's striped coralroot, 34, 44, **46**, *47*, 220, 237, 239, 242, 244, 262, 267, 278, 289, 292, 295, 307
    yellow/white form, **46**, *47*, 220

Water-spider orchid, **96**, *97*, 222, 228, 232, 236, 238, 243, 247, 252, 286, 301, 303
Western Prairie fringed orchis, 5, 129, **158**, *159*, 224, 228, 230, 234, 247, 256, 265, 267, 269, 282, 298, 301, 303–4
Western spotted coralroot, 34, 36, **38**, *39*, 219, 227, 229, 233–35, 237–38, 244, 255, 261, 267, 275, 277, 289, 292, 295, 303
    brown-stemmed form, **38**, 219
    golden yellow/spotted form, **38**, *39*, 219
    red-stemmed form, **38**, *39*, 219
    yellow spotless form, **38**, *39*, 219
White lady's-slipper, small, 5, 53, **56**, *57*, 60, 64, 220, 227, 229–30, 233–35, 237–38, 241, 246, 248, 250, 253, 283, 300, 307
White fringed orchis, northern, 129, **134**, *135*, 136, 223, 230, 233, 241, 243, 263, 266, 268, 301, 304

entire-lip form, **134**, 223
Whorled pogonia
    large, 103, **106**, *107*, 222, 228, 230, 232–33, 238, 243, 248, 250, 252, 263, 270, 286, 297, 304
    small, 103, **104**, *105*, 106, 222, 230, 233, 242–44, 263, 269, 270, 304, 307
Wilder's crested coralroot, **100**, 222
Willey's showy orchis, **78**, *79*, 221
Wister's coralroot, 34, **50**, *51*, 220, 228–29, 231–37, 286, 289, 297, 300, 303
    cranberry-stemmed form (Cooper's), **50**, *51*, 220
    red-stemmed form, **50**, *51*, 220
    white-flowered form, **50**, *51*, 220
Woodland ladies' tresses, 175, **200**, *201*, 225, 229, 232, 236, 238, 243–44, 247, 252, 256, 275, 287, 298, 303, 308
Wright's little club-spur orchis, **90**, 221

Yellow fringeless orchis, ***215***
Yellow lady's-slipper
    large, 53, 62, **64**, *65*, 220, 227–29, 231, 233, 235, 237–39, 241, 247, 254, 256, 262, 267–68, 278, 283, 286, 289, 292, 295, 297, 301
    northern small, 53, **62**, *63*, 220, 227, 229–30, 233–35, 238, 241–42, 248, 255, 267–68, 278, 283, 292, 300, 303
    southern small, 53, **60**, *61*, 64, 220, 228–29, 231, 233–34, 236, 243, 252, 255, 262, 267, 275, 278, 286, 289, 297, 303
    white-lipped form, **60**, *61*, 220

Zion, Illinois, 284

Paul Martin Brown is a research associate at the University of Florida Herbarium at the Florida Museum of Natural History in Gainesville. Originally from Massachusetts, he is a resident of Ocala, Florida, and continues to summer in Acton, Maine. He is the founder of the North American Native Orchid Alliance and editor of the *North American Native Orchid Journal*. Brown and his partner, Stan Folsom, have published *Wild Orchids of the Northeastern United States* (1997), *Wild Orchids of Florida* (UPF, 2002), *The Wild Orchids of North America, North of Mexico* (UPF, 2003), *Wild Orchids of the Southeastern United States North of Peninsula Florida* (UPF, 2004), *Wild Orchids of Florida*, revised and expanded edition (UPF, 2005), *Wild Orchids of the Canadian Maritimes and Northern Great Lakes Region* (UPF, 2005), and *Wild Orchids of the Pacific Northwest and Canadian Rockies* (UPF, 2005). *Wild Orchids of the Prairies and Great Plains Region* is their eighth volume on the wild orchids of North America. Now in preparation is *Wild Orchids of the Northeast: New England, New York, Pennsylvania, and New Jersey*.

Stan Folsom is a retired art teacher and botanical illustrator. His family is originally from Springvale and Jonesboro, Maine, and he has summered on Mousam Lake in Acton, Maine, for more than 70 years. His primary medium is watercolor, and his work is represented in several permanent collections including the Federal Reserve Bank of Boston.